Denis J. Evans, Debra J. Searles, and Stephen R. Williams

Fundamentals of Classical Statistical Thermodynamics

Denis J. Evans, Debra J. Searles, and Stephen R. Williams

Fundamentals of Classical Statistical Thermodynamics

Dissipation, Relaxation and Fluctuation Theorems

WILEY-VCH
Verlag GmbH & Co. KGaA

Authors

Prof. Denis J. Evans
Australian National University
Department of Applied Mathematics
Research School of Physics
and Engineering
Canberra ACT, 2601
Australia

Prof. Debra J. Searles
The University of Queensland
Australian Institute for Bioengineering
and Nanotechnology Centre for
Theoretical and Computational
Molecular Science
School of Chemistry and
Molecular Biosciences
Brisbane, Qld 4072
Australia

Dr. Stephen R. Williams
Australian National University
Research School of Chemistry
Building 35
Research School of Chemistry
Canberra, ACT 2601
Australia

Library of Congress Card No.: applied for

British Library Cataloguing-in-Publication Data
A catalogue record for this book is available from the British Library.

Bibliographic information published by the Deutsche Nationalbibliothek
The Deutsche Nationalbibliothek lists this publication in the Deutsche Nationalbibliografie; detailed bibliographic data are available on the Internet at <http://dnb.d-nb.de>.

Print ISBN: 978-3-527-41073-6
ePDF ISBN: 978-3-527-69576-8
ePub ISBN: 978-3-527-69577-5
Mobi ISBN: 978-3-527-69575-1
oBook ISBN: 978-3-527-69578-2

Cover Design Formgeber, Mannheim
Typesetting SPi Global, Chennai, India
Printing and Binding Markono Print Media Pte Ltd, Singapore

Printed on acid-free paper

Contents

List of Symbols

Microscopic Dynamics

N	Number of particles in the system.
D_C	Cartesian dimensions of the system – usually three.
D	Accessible phase space domain
$\mathbf{q}$	ND_C-dimensional vector, representing the particle positions.
$\mathbf{p}$	ND_C-dimensional vector, representing the particle momenta.
$\mathbf{\Gamma}$	$2ND_C$-dimensional phase space vector, representing all $\mathbf{q}$'s and $\mathbf{p}$'s.
$\delta V_{\Gamma}(S^t\mathbf{\Gamma})$	Very small volume element of phase space centered on $S^t\mathbf{\Gamma} \equiv \exp[iL(\mathbf{\Gamma})t]\mathbf{\Gamma}$.
$p(\delta V_{\Gamma}(\mathbf{\Gamma});t)$	Probability of observing sets of trajectories inside $\delta V_{\Gamma}(\mathbf{\Gamma})$ at time t.
$d(S^t\mathbf{\Gamma})$	Infinitesimal phase space volume centered on $S^t\mathbf{\Gamma}$.
$p_{+/-}(t)$	Probability that the time-integrated dissipation function is plus/minus over the time interval $(0,t)$.
M^T	Time reversal map $M^T(\mathbf{q},\mathbf{p}) \equiv (\mathbf{q},-\mathbf{p})M^T$.
M^K	Kawasaki or K-map of phase space vector for planar Couette flow, $M^K(x,y,z,p_x,p_y,p_z,\dot{\gamma}) \equiv (x,-y,z,-p_x,p_y,-p_z,\dot{\gamma})M^K$, where $\dot{\gamma} \equiv \partial u_x/\partial y$ is the xy component of the strain rate tensor.
L	f-Liouvillean.
$\exp[-i\mathsf{L}(\mathbf{\Gamma})t]\ \ldots$	f-propagator.
L	p-Liouvillean.
$\exp[iL(\mathbf{\Gamma})t]$	p-propagator.
S^t	p-propagator.
$K(\mathbf{p})$	Peculiar kinetic energy.
$\Phi(\mathbf{q})$	Interparticle potential energy.
$\phi_{i,j}(r_{ij})$	Pair potential of particle i with particle j.
$\mathbf{r}_{ij} \equiv \mathbf{r}_j - \mathbf{r}_i$	Position vector from particle i to particle j.
$r_{ij} \equiv \lvert\mathbf{r}_j - \mathbf{r}_i\rvert$	Distance between particles i and j.
$\mathbf{F}_{ij}$	Force on particle i due to particle j

$\nabla_{\mathbf{q}}$	$\equiv (\partial/\partial \mathbf{q}_1, \ldots, \partial/\partial \mathbf{q}_N$.
$H_0(\boldsymbol{\Gamma})$	Internal energy, $H_0 = K + \Phi$.
$H(\boldsymbol{\Gamma})$	Hamiltonian at phase vector $\boldsymbol{\Gamma}$.
$g(\boldsymbol{\Gamma})$	Deviation function – even in the momenta.
H_E	Extended Hamiltonian for Nosé–Hoover dynamics.
K_{th}	Peculiar kinetic energy of thermostatted particles $= D_C N_{\text{th}} k_B T_{\text{th}}/2 + O(1)$, where T_{th} is the kinetic temperate of the thermostat. If the system is isokinetic, $T_{\text{th}} = T$ – see thermodynamic variables below.
N_{th}	Number of thermostatted particles.
α	Gaussian or Nosé–Hoover thermostat or ergostat multiplier.
τ	Time constant.
$\dot{Q}_{\text{th/soi}}$	Rate of transfer of heat to the thermostat/system of interest.
Λ	Phase space expansion factor.
S_i	Switch function.
$\mathbf{J}(\boldsymbol{\Gamma})$	Dissipative flux.
$\mathbf{F}_e$	Dissipative external field.
m	Particle mass.
$\mathsf{T} \equiv \partial\dot{\boldsymbol{\Gamma}}(\boldsymbol{\Gamma})/\partial\boldsymbol{\Gamma}$	Stability matrix.
$\exp_L$	Time-ordered exponential operator, latest times to left.
$\boldsymbol{\Xi}(\boldsymbol{\Gamma}; t)$	Tangent vector propagator $\equiv \exp_L\left(\int_0^t ds\ \mathsf{T}(S^s\boldsymbol{\Gamma})\right)$.
λ_i	ith Lyapunov exponent.
$\lambda_{\text{max/min}}$	Largest/smallest Lyapunov exponent for steady or equilibrium state.

Statistical Mechanics

$\overline{A}_t$	Time average of some phase variable, $A(\boldsymbol{\Gamma})$
$\langle A(t)\rangle$	Ensemble average of A at time t, on a time-evolved path.
$f(\boldsymbol{\Gamma}; t)$	Time-dependent phase space distribution function.
$\langle\cdots\rangle_{\mu c}$	Equilibrium microcanonical ensemble average.
$\langle\cdots\rangle_c$	Equilibrium canonical ensemble average,
$f_c(\boldsymbol{\Gamma})$	Equilibrium canonical distribution.
$f_{\mu c}(\boldsymbol{\Gamma})$	Equilibrium microcanonical distribution.
Λ	Phase space expansion factor.
$\Omega(S^{t_1}\boldsymbol{\Gamma}; t_2)$	The instantaneous dissipation function, at time t_1 on a phase space trajectory that started at phase $\boldsymbol{\Gamma}$ and defined with respect to the distribution function at time t_2. $\Omega(S^{t_1}\boldsymbol{\Gamma}; 0) \equiv \Omega(S^{t_1}\boldsymbol{\Gamma})$,
$\Omega(S^{t_1}\boldsymbol{\Gamma})$	$\Omega(S^{t_1}\boldsymbol{\Gamma}) \equiv \Omega(S^{t_1}\boldsymbol{\Gamma}; 0)$
$\mathbf{r}$	Three-dimensional position vector.

$\mathbf{u}(\mathbf{r}, t)$	Three-dimensional local fluid streaming velocity, at Cartesian position $\mathbf{r}$ and time t.
$S_G(t)$	Fine-grained Gibbs entropy, $\equiv -k_B \int_D d\boldsymbol{\Gamma} f(\boldsymbol{\Gamma}; t) \ln(f(\boldsymbol{\Gamma}; t)) \overset{\text{equilibrium}}{\Rightarrow} k_B \ln(Z_{\mu c})$.
Z	Partition function – normalization for the equilibrium phase space distribution.
Z_c	Canonical partition function.
$Z_{\mu c}$	Microcanonical partition function.

Mechanical Variables

Q	Heat of thermostat.
V	Volume of system of interest.
U	Internal energy, $U = \langle H_0 \rangle$ of the system of interest.
W	Work performed on system of interest.
Y	Purely dissipative generalized dimensionless work.
X	Generalized dimensionless work.

Thermodynamic Variables

T	Equilibrium temperature the system will relax to if it is so allowed.
β	Boltzmann factor (reciprocal temperature) $\equiv 1/k_B T$.
S_{ir}	Irreversible calorimetric entropy, defined by $\Delta S_{\text{ir}} \equiv \int dt\, \dot{Q}(t)/T(t)$, where $T(t)$ is the instantaneous equilibrium temperature the system would relax to if it was so allowed. In Section 5.7, we show that the Gibbs entropy and the irreversible calorimetric entropy are equal, up to an additive constant.
$S_{\text{eq}} \equiv S$	The calorimetric entropy defined in classical thermodynamics as $\Delta S_{\text{eq}} \equiv \Delta S \equiv_{\text{qs}} \int_1^2 dt\ \dot{Q}(t)/T(t)$, where T is equilibrium temperature of the system. This entropy is a state function.
A	Helmholtz free energy; $= -k_B T \ln(Z_c) = U - TS_{\text{eq}}$.
A_{ne}	Nonequilibrium Helmholtz free energy; $= U - TS_{\text{ir}}$. This is not a state function.
$\langle \Sigma(t) \rangle$	Total entropy production – only defined in the weak field limit close to equilibrium.
G_0	Zero-frequency elastic shear modulus.
G_∞	Infinite-frequency shear modulus.

Transport

γ	Strain (note: γ is sometimes used to fix the system's total momentum).
$\delta\gamma$	Small strain.
$\dot{\gamma}$	Strain rate.
P_{xy}	xy element of the pressure tensor.
$-\langle P_{xy}\rangle$	xy element of the ensemble averaged stress tensor.
$\widetilde{\eta}_{0^+}$	Limiting zero-frequency shear viscosity of a solid.
η	Zero-frequency shear viscosity of a fluid.
τ_M	Maxwell relaxation time.
$\mathbf{J}(\mathbf{\Gamma})$	Dissipative flux.
$J_\perp(k_y, t)$	Wavector dependent transverse momentum density.
$\eta_M(t)$	Maxwell model memory function for shear viscosity.
η_M	Zero-frequency shear viscosity of a Maxwell fluid.

Mathematics

$\Theta(t)$	Heaviside step function at $t=0$.
$\forall$	For all.
$\forall!$	For almost all. The exceptions have zero measure.
λ	Arbitrary scaling parameter.
$\widetilde{F}(s)$	Laplace transform of $F(t)$.
$\widehat{F}(s)$	Anti-Laplace transform of $F(t)$.
$\oint_P$	Cyclic integral of a periodic function.
${}_{\mathrm{qs}}\int_a^b$	Quasi-static integral from a to b.
D_{KY}	Kaplan–Yorke dimension of a nonequilibrium steady state.

Note: Upper case subscripts/superscripts indicate people. Lower case is used in most other cases. Subscripts are preferred to superscripts so as to not confuse powers with superscripts. Italics are used for algebraic initials. Nonitalics for word initials. (*e.g.*, T-mixing not T-mixing because T stands for Transient, N-particle not N-particle.)

Acknowledgments

DJE, DJS, and SRW would like to thank the Australian Research Council for the long-term support of the research projects that ultimately led to the writing of this book. DJE would also like to acknowledge the assistance of his many fourth-year honors, science, and mathematics students who over the last decade took his course: "The Mathematical Foundations of Statistical Thermodynamics." The authors also wish to thank their former PhD students for their contributions to some of the material described in this book: especially, Dr Charlotte Petersen, Dr Owen Jepps, Dr James Reid, and Dr David Carberry. The first experimental verification of a fluctuation theorem was carried out in Professor Edith Sevick's laboratory at ANU with excellent technical support provided by Dr Genmaio Wang. The authors would also like to thank Professor Lamberto Rondoni who helped them understand some of the elements of ergodic theory.

1 Introduction

> The instantaneous reversal of the motion of every moving particle of a system causes the system to move backwards, each particle along its path and at the same speed as before …
>
> *(Thomson, 1874)*

Until very recently, the foundations of statistical mechanics were far from satisfactory (Evans, Searles, and Williams, 2009a). Textbooks approach the derivation of the canonical distribution in one of two ways. A common approach is to *postulate* a microscopic definition for the entropy and then to show that the standard canonical distribution function can be obtained by maximizing the entropy subject to the constraints that the distribution function should be normalized and that the average energy is constant. The choice of the second constraint is completely subjective due to the fact that, at equilibrium, the average of every phase function is fixed. The choice of the microscopic expression for the entropy is also *ad hoc.* This "derivation" is therefore flawed.

The second approach begins with Boltzmann's *postulate* of equal *a priori* probability in phase space for the microcanonical ensemble and then derives an expression for the most probable distribution of states in a small subsystem within that much larger microcanonical system. A variation of this approach is to simply *postulate* a microscopic expression for the Helmholtz free energy via the partition function.

The so-called Loschmidt paradox, which so puzzled Boltzmann and his contemporaries, remained unresolved for 119 years after it was first raised. All the equations of motion in mechanics (both classical and quantum) and electrodynamics are time-reversal-symmetric. Time reversibility of the classical equations of motion is trivial to demonstrate. Consider Newton's equations of motion for the positions $\mathbf{q}_i$ of N identical particles subject to interatomic forces $\mathbf{F}_i(\mathbf{q}_1, \ldots, \mathbf{q}_N)$:

$$m\frac{d^2\mathbf{q}_i(t)}{dt^2} = \mathbf{F}_i(\mathbf{q}), \quad i = 1, \ldots, N \tag{1.1}$$

As Loschmidt and Kelvin (separately) noticed (Loschmidt, 1876; Thomson, 1874), time reversal $t \to -t$ leaves Eq. (1.1) unaltered since $(-1)^2 = 1$. This means that if $\mathbf{q}(t); -\tau < t < \tau$ is a solution of the equations of motion, then so too is

Fundamentals of Classical Statistical Thermodynamics: Dissipation, Relaxation and Fluctuation Theorems, First Edition. Denis J. Evans, Debra J. Searles, and Stephen R. Williams.

$\mathbf{q}(-t): \quad -\tau < t < \tau$. Changing the direction of time inverts every velocity – as per Kelvin's quote above.

The Loschmidt Paradox can be stated quite simply. If all the laws of physics are time-reversal-symmetric, how can one prove a time-asymmetric law like the second "Law" of thermodynamics that states that the entropy of the Universe "tends to a maximum" (Clausius, 1865; Clausius, 1872). Although there have been many attempts over the last century to resolve this paradox, the matter was not really settled until the first proof of a fluctuation theorem in 1994 (Evans and Searles, 1994).

A less well-known problem concerns Clausius' inequality itself. In some ways, this is an even more fundamental problem because it concerns thermodynamics rather than statistical mechanics. Clausius' inequality for the heat Q_{th} transferred to a thermal reservoir states that the cyclic integral $\oint dQ_{\text{th}}/T \geq 0$. When this inequality is, in fact, an *equality* (the process is *quasi-static*), we have the usual argument that $\int dQ_{\text{th}}/T_{\text{th}}$ is a state function and represents the change in the equilibrium entropy of the reservoir, S_{th} and T_{th} is the equilibrium thermodynamic temperature of that reservoir or set of reservoirs. Clausius went on to apply his inequality to the system of interest (soi) and thermal reservoir (th). Indeed, in his original papers he does not distinguish between the two systems.

Now comes the difficulty: when we have a strict inequality $\oint dQ/T > 0$, either the system of interest or the reservoir (or both) is (or are) not in true thermodynamic equilibrium (the process is not *quasi-static*). In this case, what is the temperature? Clausius only defined the temperature for quasi-static or equilibrium processes where the entropy is a state function. In the case of a strict inequality, $\int dQ/T$ is *not* a state function. It is path- and/or history-dependent.

For quasi-static processes (only!), the change in equilibrium entropies of two equilibrium states can be obtained by considering $\int dQ_{\text{th}}/T_{\text{th}}$ for a reversible (i.e., infinitely slow) pathway between the two equilibrium states. However, if the initial or final states are out of equilibrium or if the pathway connecting the two states is irreversible, the entropy that Clausius defined is ill-defined and so too is the temperature: $T \equiv \partial U/\partial S|_V$, where U is the internal energy, S the (undefined) entropy, and V the volume. This means that the Clausius *inequality* $\oint dQ/T > 0$ is without meaning.

Clausius is famous for his declaration:

> The energy of the Universe is constant. The entropy tends to a maximum.
> *(Clausius, 1865, 1872)*

He did not recognize the fact that he only defined the entropy (and temperature) for reversible processes. This particular difficulty was first discussed in the late nineteenth century by Bertrand (1887) and early in the twentieth century by Orr (1904), Orr (1905), Planck (1905), and Buckingham (1905).

"There are three things in Prof. Orr's article (Orr, 1904) which stand out as of particular importance. (1) He says in substance, though with great moderation, that all proofs of the theorem … when the integral is taken round an irreversible cycle, are rubbish." Buckingham later discusses problems with writing textbooks while being aware at the time, of some of the difficulties mentioned above. Buckingham continues: "The question how a treatise should be written is not so easily answered. … I do not know of a single book which today deserves the title of 'Treatise on Thermodynamics'." He concluded: "We must leave the question of the proper method for a treatise to the future when the difficulties which now beset us may have vanished." (Buckingham, 1905)

In 1905, Planck responded to Orr (Planck, 1905) agreeing with Orr's concerns on the definition of temperature and saying in part that: "If a process takes place so violently that one can no longer define temperature …, then the usual definition of entropy is inapplicable."

These particular difficulties were only exacerbated in 1902 with the publication (and subsequent circulation) of Gibbs' seminal treatise "Elementary Principles in Statistical Mechanics" (Gibbs, 1981). In his treatise, Gibbs showed that the microscopic expression he identified at equilibrium, as the thermodynamic entropy $S_G(t) \equiv -k_B \int d\mathbf{\Gamma} f(\mathbf{\Gamma};t) \ln[f(\mathbf{\Gamma};t)]$, where $f(\mathbf{\Gamma};t)$ is the N-particle phase space distribution function at time t, is in fact a constant of the motion for autonomous Hamiltonian dynamics! If the initial distribution was not the equilibrium distribution, the Gibbs entropy did not, as Clausius claimed, increase in time until it reached its maximum and the system was effectively in equilibrium. For these systems, the Gibbs' entropy is simply a constant independent of time.

After Boltzmann's death, this distressing state of affairs was reviewed without satisfactory resolution by the Ehrenfests in 1911 (Ehrenfest and Ehrenfest, 1990). (Paul Ehrenfest was a student of Boltzmann.) Indeed ,in the Preface to the (English) Translation, Tatiana Ehrenfest confides: "At the time the article was written [1911], most physicists were still under the spell of the derivation by Clausius of the existence of an integrating factor for the … heat … it became clear to me afterwards, that the existence of an integrating factor has to do only with the differentials $dx_1, dx_2, \ldots, dx_n$ of the *equilibrium* [T. Ehrenfest's italics] parameters $dx_1, dx_2, \ldots, dx_n$, and is completely independent of the direction of time … Nevertheless even today [1959] many physicists are still following Clausius, and for them the second law of thermodynamics is still identical with the statement that entropy can only increase."

The Ehrenfests' article did point out that away from equilibrium entropy was problematic and that for autonomous Hamiltonian systems the entropy defined by Gibbs was indeed a constant of the motion. In Ehrenfest and Ehrenfest (1990, p. 54), they agree with Gibbs that, "From Liouville's theorem, Eqs. (26) and (26'), it follows immediately that the quantity σ [i.e., S_G above] … remains exactly constant during the mixing process." They go on to discuss Gibbs' flawed attempts to resolve the paradox by defining a coarse-grained entropy. This quantity's time dependence is determined by the grain size and is thus not an objective property of the physical system of interest.

The theory of the relaxation to equilibrium has also been fraught with difficulties (Evans, Searles, and Williams, 2009a). First, there was no mathematical definition of equilibrium! The first reasonably general approach to this problem is summarized in the Boltzmann H-theorem. Beginning with the definition of the H-function, Boltzmann proved that the Boltzmann equation for the time evolution of the single particle probability density implies, for uniform ideal gases, a monotonic decrease of the H-function in time (Boltzmann, 1872) – see the review by Lebowitz (1993) for a modern discussion of Boltzmann's ideas.

However, there are at least two problems with Boltzmann's treatment. First, the Boltzmann equation is valid only for an ideal gas. Second, and more problematic, unlike Newton's equations, Hamilton's principle, or the time-dependent Schrödinger equation, the Boltzmann equation itself is *not* time-reversal-symmetric. It is therefore completely unsurprising that the Boltzmann equation predicts a time-irreversible result, namely the Boltzmann H-theorem.

This leads to a second version of the irreversibility paradox (at least for ideal gases): how can the time-irreversible Boltzmann equation, which leads easily to the time irreversible Boltzmann H-theorem, be derived exactly for ideal gases from time-reversible Newton's equations? This issue was also discussed, without resolution, in the Ehrenfest encyclopedia article (Ehrenfest and Ehrenfest, 1990).

Since our new proof of how macroscopic irreversibility arises from time-reversible microscopic dynamics is valid for all densities, we do not need to directly answer this question in this book. We do make the comment, however, that it is thought that in the ideal gas limit, the Boltzmann equation is exact, but its detailed derivation is beyond the scope of this present book.[1)]

The 1930s saw significant progress in ergodic theory with a proof that for a finite, autonomous Hamiltonian system, whose dynamics preserves a *mixing* microcanonical equilibrium distribution (i.e., a distribution that is uniform over the constant energy phase space hypersurface), averages of physical properties must, in the long-time limit, approach those obtained with respect to that equilibrium microcanonical distribution, regardless of the initial distribution (Sinai, 1976). Later in this book we will give a generalization of the ergodic theory proof. We consider finite systems with autonomous dynamics that are mixing with respect to some possibly thermostatted and/or barostatted equilibrium distribution that is also a solution to the dynamics considered. We show that for such systems, at sufficiently long times, averages of physical phase functions will approach, to arbitrary accuracy, the equilibrium averages taken over their mixing equilibrium distributions, irrespective of the initial distribution.

These proofs are, however, not very revealing. They tell us almost nothing of the relaxation process, only that it takes place. Relaxation is inferred rather than elucidated.

We go on to discuss a new set of theorems and results that, when taken together, provide a completely new approach to establishing the foundations of classical statistical thermodynamics and simultaneously resolving all the issues mentioned above. Each of these theorems is consistent with time-reversible,

1) In Chapter 9, we do make some comments on the relationship between Boltzmann's assumption of molecular chaos (stosszahlansatz in German) and the axiom of causality. It is this assumption that breaks time reversal symmetry in the Boltzmann equation.

deterministic dynamics. Indeed, time reversibility of the underlying equations of motion is the key component to proving these theorems. We do comment that there are stochastic and/or quantum versions of some of the theorems. Each of these theorems is exact for systems of arbitrary size: taking the thermodynamic limit is not required. The theorems are valid for arbitrary temperatures and densities. The theorems are exact arbitrarily near to, or far from, equilibrium. Assumptions about being arbitrarily close to equilibrium, so that the response of systems to external forces is linear, are not required. In the process of deriving these theorems, the so-called "Laws" of thermodynamics cease to be unprovable "Laws" and instead become mathematical theorems.

The first step toward understanding how macroscopic irreversibility arises from microscopically time-reversible dynamics came in 1993 when Evans, Cohen, and Morriss (1993) proposed the first so-called fluctuation relation. By generalizing concepts from the theory of unstable periodic orbits in low-dimensional systems, these authors proposed a heuristic, asymptotic argument for the relative probability of seeing sets of trajectories and their conjugate sets of antitrajectories in nonequilibrium steady states maintained at constant internal energy. In the following year, Evans and Searles (1994) published the first mathematical proof of a fluctuation theorem. A generalized and detailed proof of the Evans–Searles fluctuation theorem is given in Chapter 3. This proof concerns the relative probability of fluctuations in sign of a quantity now known as the time-averaged *dissipation function*. Unsurprisingly, fluctuation theorems lead to many new results. This is what the present book sets out to describe. It used to be said that there are very few exact results that are known for nonequilibrium many-body systems. This is no longer the case.

In Chapter 3, we prove the second law inequality (Searles and Evans, 2004), and the nonequilibrium partition identity (Morriss and Evans, 1985; Carberry *et al.*, 2004; Evans and Searles, 1995). These are simple mathematical consequences of the fluctuation theorem. The second law inequality is, in fact, a generalization of the second "Law" of thermodynamics that is valid for finite, even small systems, observed for finite, even short, times. Classical thermodynamics applies to only large, in principle infinite, systems either at equilibrium or in the infinitely slow, or quasi-static, limit.

Dissipation was first explicitly defined in 2000 by Searles and Evans (2000a), although it was, of course, implicit in the earlier proofs of the Evans–Searles fluctuation theorems in 1994, et seq. It is also implicit in many of Lord Kelvin's papers in the late nineteenth century. The dissipation function has many properties, but its original definition directly involved sets of trajectories and their conjugate sets of time-reversed antitrajectories. For classical N-particle systems, the specification of all the coordinates and momenta of all the atoms in the system completely describes the microstate of a classical system. We define the phase space vector $\mathbf{\Gamma} = (\mathbf{q}_1, \ldots, \mathbf{q}_N, \mathbf{p}_1, \ldots, \mathbf{p}_N)$ of the positions $\mathbf{q}_i$ and momenta $\mathbf{p}_i$ of the N particles. We imagine an infinitesimal set of phases inside an infinitesimal volume $dV_{\Gamma}(\mathbf{\Gamma})$ in phase space. For simplicity, we assume that the system is autonomous (i.e., the equations of motion for all the particles, $\dot{\mathbf{\Gamma}}(\mathbf{\Gamma}, t)$, do not refer explicitly to time $\dot{\mathbf{\Gamma}}(\mathbf{\Gamma})$; any external fields are time-independent).

As time evolves, this set will trace out an infinitesimal tube in phase space. We follow this tube for a time interval $(0, t)$. At time t, an initial phase space vector $\mathbf{\Gamma}$ has evolved to the position $S^t\mathbf{\Gamma}$, where S^t is the phase space–time evolution operator. If we take the set of phase points inside the infinitesimal volume $dV_{\mathbf{\Gamma}}(S^t\mathbf{\Gamma})$ and reverse all the momenta leaving all the particle positions unchanged, we have the phase vector $M^T S^t\mathbf{\Gamma}$, where M^T is a time-reversal mapping: $M^T(\mathbf{q}, \mathbf{p}) = (\mathbf{q}, -\mathbf{p})$.

If we now imagine following the natural motion of this mapped set forward in time from time t to $2t$, we arrive at the phase point $S^t M^T S^t\mathbf{\Gamma}$. Because the equations of motion are time-reversal-symmetric, the final set of phase points will have the same position coordinates but the opposite momenta to the original set of time zero phases: $S^t M^T S^t\mathbf{\Gamma} = M^T\mathbf{\Gamma}$. This is the fundamental property of time-reversible dynamics discussed in Kelvin's quote at the beginning of this chapter. This time reversibility property is exploited directly in the definition of the dissipation function. We will give a more detailed description of reversibility using a more precise notation in Chapter 2 – especially in Section 2.1.

The time integral of the dissipation function is simply defined as the natural logarithm of the probability ratio of observing at time zero the conjugate sets of trajectories inside phase space volumes $\delta V_{\mathbf{\Gamma}}(\mathbf{\Gamma})$, $\delta V_{\mathbf{\Gamma}}(M^T S^t\mathbf{\Gamma})$:

$$\lim_{\delta V_{\Gamma}\to 0} \frac{p(\delta V_{\mathbf{\Gamma}}(\mathbf{\Gamma}); 0)}{p(\delta V_{\mathbf{\Gamma}}(M^T S^t\mathbf{\Gamma}); 0)} \equiv \exp\left[\int_0^t ds\,\Omega\,(S^s\mathbf{\Gamma})\right] \tag{1.2}$$

The small phase space volume $\delta V_{\mathbf{\Gamma}}(\mathbf{\Gamma})$ defines an initial set of phase space trajectories. The volume $\delta V_{\mathbf{\Gamma}}(M^T S^t\mathbf{\Gamma})$ defines the conjugate set of the antitrajectories. Going forward in time from $\delta V_{\mathbf{\Gamma}}(M^T S^t\mathbf{\Gamma})$ is like going backward in time from $\delta V_{\mathbf{\Gamma}}(S^t\mathbf{\Gamma})$ except that all the momenta are reversed. For Eq. (1.2) to be well defined requires that the system should be *ergodically consistent*, that is, if the numerator is nonzero for initial phases inside some specified phase space domain D, then the denominator must also be nonzero. This condition ensures that the dissipation function is well defined *everywhere* inside the ostensible phase space domain, D.

As a historical remark, we can see from the definition, Eq. (1.2), that ergodic consistency guarantees the existence of (almost all) conjugate phase space trajectory/antitrajectory pairs. However, the mere existence of these pairs of trajectories by no means implies that the probability ratio of observing infinitesimal *sets* of these conjugate trajectory pairs is unity, as Loschmidt tried to imply. Once you have written down Eq. (1.2) for the relative probability of seeing a set of trajectories and its conjugate set of antitrajectories, it seems obvious that Loschmidt's assertion of both sides of Eq. (1.2) equaling unity is wrong. One must not make the mistake of discussing *individual* conjugate phase space trajectories rather than conjugate *sets* of trajectories. The probability of observing any individual phase space trajectory is precisely zero! Their rephrasing of Eq. (1.2) would have been ill defined, leading to zero divided by zero on the left-hand side.

We will see in Chapter 5 that an equilibrium state is characterized by a set of equations of motion and a phase space distribution for which the dissipation function is identically zero everywhere in phase space. Thus, for equilibrium states alone the probabilities of observing every infinitesimal set of trajectories and its conjugate set of antitrajectories are identical. Loschmidt's assertion is correct only for equilibrium distributions. Indeed, this gives statistical thermodynamics, for the first time, a mathematical definition of an equilibrium system.

Although the definition of the dissipation function may appear rather abstract and mathematical, it turns out that in the linear regime close to equilibrium the average of the dissipation function is equal to a quantity that is familiar in linear, irreversible thermodynamics, namely the spontaneous entropy production. For systems that are driven by an applied dissipative field (e.g., an electrically conducting system being driven by an electric field), the average dissipation is equal to the average power dissipated in the system divided by the thermodynamic temperature of the surrounding thermal reservoir to which the dissipated work, on average, eventually relaxes. A notable aspect of our exposition is the fact that except at equilibrium, entropy plays no role. This neatly bypasses the objections of Bertrand, Orr, and Buckingham to the Clausius inequality for non-quasi-static processes.

The first theorem that referred to dissipation was the Evans–Searles fluctuation theorem (Evans and Searles, 1994) (FT). This theorem considers systems with time-reversible dynamics where the initial distribution of phases is even in the momenta and which satisfies the condition of ergodic consistency. It states that for such systems the ratio of probabilities that the time-averaged dissipation function $\overline{\Omega}_t$ takes on an arbitrary value in the range $A \pm dA$, compared to the negative of that value $-A \pm dA$ satisfies the following equation

$$\frac{p(\overline{\Omega}_t = A)}{p(\overline{\Omega}_t = -A)} = e^{At} \tag{1.3}$$

where $p(\overline{\Omega}_t = A)$ represents the ratio of probabilities that the time-averaged dissipation function $\overline{\Omega}_t$ takes on an arbitrary value in the range $A \pm dA$. This shows that the probability of positive dissipation is exponentially more likely than negative dissipation and, moreover, the argument of the exponential is extensive in both the number of particles in the system N and the averaging time t. Equation (1.3) has been confirmed both by molecular dynamics computer simulations and in actual laboratory experiments. The first unambiguous laboratory demonstration of a fluctuation relation was conducted in 2002 using a colloidal suspension and optical tweezers (Wang *et al.*, 2002).

A trivial consequence of the FT is the second law inequality, which states that, if we average the response of repeated experiments on our system with macroscopically identical initial conditions, the so-called ensemble average of the time-averaged dissipation $\langle\overline{\Omega}_t\rangle$ is nonnegative:

$$\langle\overline{\Omega}_t\rangle \geq 0, \quad \forall t \tag{1.4}$$

This does not imply that the instantaneous ensemble-averaged dissipation is nonnegative. This ensemble-averaged *instantaneous* dissipation $\langle\Omega(t)\rangle$ may be positive or negative, but it is, of course, positive more often than it is negative in order to satisfy Eq. (1.4).

The second law inequality also shows that

$$\Omega(\mathbf{\Gamma}) = 0, \quad \forall \mathbf{\Gamma} \in D \Leftrightarrow \langle \overline{\Omega}_t \rangle = 0 \tag{1.5}$$

The proof is rather straightforward. Obviously, the left-hand side implies the right. Does the right imply the left? Suppose the ensemble-averaged time integral of the dissipation is not identically zero everywhere. Average the dissipation over some possibly short time interval $(0,t)$. Ergodic consistency implies the existence of conjugate sets of trajectories with opposite values for the time-averaged dissipation $\pm(A + dA)$. Applying the FT to each conjugate set with time-averaged dissipation $\pm(A + dA)$ shows that positive dissipation is exponentially more likely than negative for each value of $|A|$ that is observed. If we now average over all possible values for $|A|$ for which there is nonzero dissipation, we see that $\langle \Omega_t \rangle > 0$. For any nonequilibrium system, the ensemble average of the time-integrated dissipation must be strictly positive. So, if the dissipation is nonzero anywhere in the allowed phase space and the system is ergodically consistent, then the time-averaged, ensemble-average dissipation must be strictly positive. The only states where the ensemble-averaged, time-averaged dissipation is zero are equilibrium states where the instantaneous dissipation is identically zero everywhere in the allowed phase space.

The recently discovered dissipation theorem (Evans, Searles, and Williams, 2008a,b) (Chapter 4) states that the ensemble average of an arbitrary, integrable function of phase $B(\mathbf{\Gamma})$ is related to the time integrals of the correlation function of the dissipation function with the phase variable:

$$\langle B(t) \rangle = \langle B(0) \rangle + \int_0^t ds \langle B(s)\Omega(0) \rangle \tag{1.6}$$

The dynamics employed for evaluating *all* functions on both sides of Eq. (1.6) employs natural system dynamics including any external fields and/or thermostats. This result is valid arbitrarily far from equilibrium and for systems of arbitrary size. In systems where an externally applied field is responsible for driving the system out of equilibrium in the weak field regime where the response to this field is linear, Eq. (1.6) reduces to the very well known Green–Kubo linear response equations (Evans and Morriss, 1990).

Since the instantaneous average dissipation is zero for equilibrium systems, Eq. (1.6) shows that, in the absence of an external field, ensemble averages of phase function never change for systems at equilibrium. It turns out that for equilibrium systems the equilibrium distribution itself never changes.

Together with the definition of dissipation, a second very important definition is that of an ΩT-mixing system. A system is said to be ΩT-mixing if infinite time integrals of ensemble averages of phase variables $B(\mathbf{\Gamma})$, representing physical observables like pressure, stress, energy, and so on, multiplied by the dissipation function and evaluated at time zero are convergent: $(\lim_{t\to\infty} \left| \int_0^t ds \langle B(s)\,\Omega(0) \rangle \right| = \text{const} < \infty)$. A system of harmonic oscillators with zero friction is obviously not ΩT-mixing. ΩT-mixing is a more physically relevant condition than the mixing condition met in ergodic theory. From Eq. (1.6), we see that, if an autonomous system is ΩT-mixing, then at long times the ensemble average of physical phase functions become time-independent at long times. At long times, ΩT-mixing systems must therefore relax either toward nonequilibrium *steady* states or toward equilibrium states. No other possibilities exist.

If the infinite time integral of ensemble averages of time correlation functions of physical phase functions all $A(\mathbf{\Gamma})$ and $B(\mathbf{\Gamma})$ is finite (i.e., $\lim_{t\to\infty}\left|\int_0^t ds\,\langle A\,(0)\,B(s)\rangle\right| = \text{const} < \infty$) when the ensemble average of $A(\mathbf{\Gamma})$ is zero, that is, $\langle A(\mathbf{\Gamma})\rangle = 0$, then the system is termed *T-mixing*. Obviously all T-mixing systems are ΩT-mixing. ΩT-mixing systems are not necessarily T-mixing. Note that any phase function with a nonzero ensemble average (say $\widetilde{A}(\mathbf{\Gamma})$) can be transformed into one with zero average, $\widetilde{A}(\mathbf{\Gamma}) - \langle\widetilde{A}(\mathbf{\Gamma})\rangle = A(\mathbf{\Gamma})$.

The dissipation function, ergodic consistency, and the T-mixing condition hold over some specified phase space domain D. For example, while particle momenta may be unbounded, the particle coordinates are usually defined only over a fixed region of the physical space. A system is said to be *physically ergodic* over some specified phase space domain if time averages of phase functions representing physical observables taken along almost any phase space trajectory equal late-time ensemble averages taken over any ensemble of initial states.

T-mixing systems must be physically ergodic over that specified phase space domain. If they were not, we could easily construct time correlation functions of physical observables that would never decay to zero. Any initial static correlation between the phase functions would be preserved forever, thereby violating the condition of T-mixing.

Physically, ergodic systems need not be *ergodic over phase space.* Different initial phase space vectors generate, via their different trajectories, different nonintersecting sets of phase space subdomains – one subdomain corresponding to each phase space trajectory and parameterized by time $(0, \infty)$. If the time average of physical properties along each of the different trajectories is independent of the particular trajectory, the system may be *physically ergodic* but not *ergodic over phase space.* This could occur because each trajectory shadows the other trajectories in a densely woven “mat.” In this book, we will deal almost exclusively with *physical ergodicity,* which we will refer to simply as *ergodicity*. On the rare occasions that we refer to *ergodicity over phase space,* we will make that explicit at the time. Of course, if a system is *ergodic over phase space,* it must also be *physically ergodic.*

The equilibrium relaxation theorem (Evans, Searles, and Williams, 2009a,b) derived in Chapter 5 states that autonomous N-particle T-mixing systems that may be isolated or perhaps interact with a heat bath and whose initial distributions are even functions of the momenta will, at sufficiently long times, relax toward a unique equilibrium state and that

$$\lim_{t\to\infty}\langle\Omega(S^t\mathbf{\Gamma})\rangle = 0, \quad \forall\mathbf{\Gamma}\in D \tag{1.7}$$

For various forms of thermostat or ergostat, the unique forms of these equilibrium distributions can be determined explicitly using the various individual forms of the equilibrium relaxation theorem. Since *any* reasonably smooth deviation from the unique equilibrium dissipation causes the ensemble-averaged dissipation to be positive, the only conclusion from Eq. (1.7) is that, in the infinite time limit, the system apparently relaxes to its unique equilibrium distribution.

For constant energy dynamics, the equilibrium distribution is uniform over the energy hypersurface[2)] in phase space. The equilibrium relaxation theorem therefore gives a proof of Boltzmann's postulate of equal *a priori* probability for constant energy systems. The relaxation theorem does not imply that all relaxation processes are monotonic in time (i.e., averages of phase functions change monotonically). This is just as well, since experience shows that most relaxation processes are *not* monotonic. For thermostatted systems where the number of particles and the volume are fixed, the unique equilibrium distribution is the well-known canonical distribution postulated by Boltzmann and Gibbs.

An interesting result that we obtain from the equilibrium relaxation theorems is that relaxation to equilibrium *cannot* take place in finite time. In a sense, the equilibrium *distribution* is never reached. It is only *averages* of physical properties that approach, in the infinite time limit, the values one would obtain from a true equilibrium distribution. The actual time dependent phase space distribution becomes, at long times, ever more tightly folded upon itself. It never *becomes* a smooth equilibrium distribution. However, as the equilibrium relaxation theorems prove, the ensemble-averaged dissipation does go to zero in the infinite time limit and in that infinite time limit the distribution must be the unique smooth equilibrium distribution at least as can be ascertained by computing averages of physical phase functions like the dissipation function.

Having determined the equilibrium distribution for systems in contact with a heat reservoir, we show that the standard expression for the change in the calorimetric entropy of the system of interest, $\Delta S_{\mathrm{soi}} = \int dQ_{\mathrm{soi}}/T$, where dQ_{soi} is the change in the heat added to the system of interest, is, in fact, for quasi-static processes (processes carried out in the infinitely slow limit) a path- and history-independent state function. We show that the so-called integrating factor for the heat, namely $1/T$, which generates the corresponding state function, is in fact unique. No other integrating factor (e.g., $1/T^3$) can generate a state function from the heat. The integrating factor comes directly from the form of the equilibrium canonical distribution function, which is itself unique.

For macroscopic systems, we also derive the fundamental equation for the first and second "laws" of thermodynamics. This equation relates changes in the internal energy U to the equilibrium temperature T appearing in the equilibrium phase space distribution function, the change in the calorimetric entropy, the mechanical pressure p, and the change in the volume dV:

$$dU = TdS - pdV \tag{1.8}$$

2) The "hypersurface" is defined as $\lim_{\delta E \to 0} \{\boldsymbol{\Gamma} : E < H(\boldsymbol{\Gamma}) < E + \delta E\}$.

In Eq. (1.8), all quantities are for the system of interest. This macroscopic result is obtained entirely from microscopic or molecular expressions for the various variables.

We also show the identity (up to an arbitrary additive constant) of the Gibbs entropy and the newly defined irreversible calorimetric entropy. The equivalence of changes in the Gibbs and irreversible calorimetric entropies is valid even for irreversible processes where (and unlike Clausius) we take the temperature at any point in a process to be the equilibrium thermodynamic temperature the system would relax to if it was so allowed. The nonequilibrium temperature is, in fact, the equilibrium thermodynamic temperature of the *underlying equilibrium state* toward which the nonequilibrium system is trying to relax.

The derivation of Eq. (1.8) for quasi-static processes (only) is completely consistent with Tatiana Ehrenfest's statement quoted above that, effectively, Eq. (1.8) is "completely independent of the direction of time" (Ehrenfest and Ehrenfest, 1990).

In Chapter 6, we discuss the steady-state relaxation theorem. For systems that are initially in equilibrium for the zero-field dynamics, if a dissipative field is then applied to the system and it is T-mixing, the system will eventually relax to a physically ergodic, nonequilibrium steady state. At long times, time averages of physical phase functions equal late-time ensemble averages. Further we will show that, if the initial equilibrium distribution is perturbed by some reasonably smooth deviation function (even in the particle momenta), the final steady state is independent of the initial perturbation.

Also in Chapter 6, we discuss asymptotic steady-state fluctuation theorems (Searles and Evans, 2000b; Williams, Searles, and Evans, 2006; Searles, Rondoni, and Evans, 2007). For T-mixing systems, these steady-state fluctuation relations are valid even for large deviations from the mean behavior of the system.

In Chapter 7, we describe more theoretical applications of the fluctuation, dissipation, and relaxation theorems. A proof is given of the zeroth law of thermodynamics (Evans, Williams, and Rondoni, 2012); a discussion is given of heat flow and (Evans, Searles, and Williams, 2010) temperature quenches from the point of view of nonequilibrium statistical mechanics. A discussion is given on the relaxation of a color field gradient in a system where the Hamiltonian is color blind. In the linear response regime, as far as its Hamiltonian can sense, the system is in equilibrium. Finally, we give a derivation of an instantaneous fluctuation theorem (Petersen, Evans, and Williams, 2013).

In Chapter 8, we discuss the Crooks fluctuation relations (Crooks, 1998) and the Jarzynski equality (Jarzynski, 1997). These relations show how equilibrium free energy differences can be computed from nonequilibrium path integrals of the work. Using various generalizations of these relations we give a mathematical proof of Clausius' inequality for thermal reservoirs in contact with our system of interest. We consider a set of large thermal reservoirs at a set of temperatures. Because the reservoirs are large compared to the system of interest, they can be regarded as being in thermodynamic equilibrium. We prove (Evans, Williams, and Searles, 2011) for systems that have a periodic response to some cyclic protocol, the ensemble average of the cyclic time integral of the heat transferred to the reservoirs divided by the corresponding reservoir temperature is nonnegative. Clausius proved his inequality by *assuming* the second law of thermodynamics – the impossibility of constructing a perpetual motion machine of the second kind. Our proof makes no such assumption. Since Clausius' inequality is often taken as the most fundamental statement of the second law, our proof constitutes a direct proof of this statement of the second "Law." We show that it is true only if the system responds periodically to the cyclic protocol (not all systems do this of course), and it is true only if we take the ensemble-averaged response. A single cycle for an individual system, if it is small, may not satisfy Clausius' inequality as it applies to the reservoir.

We also show that, if the reservoirs are small and cannot be regarded as being in thermodynamic equilibrium, the ensemble average of the cyclic integral for the reservoir still satisfies Clausius' inequality. Of course, it only applies if the system responds periodically. At each point in the cycle, the temperature appearing in our generalization of Clausius' inequality is the equilibrium temperature that the entire system would relax to, if the execution of the protocol is stopped and the entire system is allowed to relax to equilibrium.

An immediate consequence of our proof of Clausius inequality for the reservoir is that the change in the entropy of the "universe": $dQ_{\text{th}}/T_{\text{th}} + dQ_{\text{soi}}/T_{\text{soi}} = 0$, where "soi" denotes the system of interest, which, by construction, is in thermal contact with the thermal reservoir "th," and is precisely zero. This result is valid for both quasi-static and nonequilibrium processes far from equilibrium using the irreversible calorimetric definition of the entropy. Since we have already proved the equivalence of changes in the irreversible calorimetric and Gibbs entropies, this new result is consistent with the observation made by Gibbs that the Gibbs entropy for an autonomous Hamiltonian system is a constant of the motion. This, of course, contradicts the claim by Clausius that the entropy of the "Universe" tends to a maximum. Furthermore, because we give meaning to temperature far from equilibrium, unlike Clausius' original inequality our result is well defined away from equilibrium and is immune to the criticisms made by Bertrand (1887), Orr (1904), and Buckingham (1905) of the original Clausius *inequality*.

Entropy and dissipation are thus seen to be completely complementary. Away from equilibrium, dissipation is the function that is central to all the theoretical results while entropy plays only a trivial roll. At equilibrium or in the quasi-static limit, dissipation is zero *by definition*, while entropy is one of the key quantities in equilibrium thermodynamics.

In Chapter 9, we revisit the proof of the Evans–Searles FT and discuss the role played therein by the axiom of causality (Evans and Searles, 1996). We prove that in an anti-causal Universe there is an anti-second "Law" of thermodynamics and that ultimately the explanation for the macroscopic irreversibility we see around us is causality. In very few discussions of irreversibility is it realized that, if you apply a time-reversal mapping to a system trajectory, not only do you reverse the direction of the flow of heat and work but the causal response to some time-dependent field becomes anti-causal! Fluxes respond to changes in field strength *before* those changes occur!

If we watch a movie played backwards of macroscopic machines in motion, not only will we see examples of "perpetual motion machines of the second kind" but we will also see a Universe where effect *precedes* the cause. The transient response to a sudden application of a cause will have the opposite sign to that observed in the forward movie, but that transient response will start *before* the change in the cause has actually occurred!

For example, in a viscometer that is loaded with a viscoelastic fluid, the shear stress in an anti-causal Universe not only has the opposite sign to that which it has in our causal Universe but it will begin to respond (negatively) *before* a shear rate is applied. Likewise, it will begin to decrease towards zero before, not after, the strain rate has been set to zero!

In a causal Universe, one needs to compute the probabilities of events occurring at a time t from the probabilities of prior events and not from the probabilities of events at times later than t. This assumption of causality breaks the time reversal symmetry of the whole system while retaining time-reversible equations of motion.

The assumption of causality seems so ingrained and natural to the human way of thinking that we often do not realize that it is an assumption. It is this assumption, or rather it is the use of this axiom in the proofs of the Evans–Searles and Crooks FTs, that breaks the symmetry of time and leads to the second law inequality rather than an anti-second law inequality.

The principle of least action, which is completely time-reversal-symmetric, does not contain sufficient information to prove any fluctuation theorem. The equations of motion of mechanics must be supplemented with the axiom of causality to predict the operation of machines, engines, and devices in the real world. The axiom is constantly being applied without us even noticing, precisely because it seems so natural. The response of a system (engine) at a given time is obtained by convolving the response function for the system with the time-dependent driving force backward over the past history and *not* over its future. The underlying equations of motion themselves retain their time reversal symmetry.

A clear example of the unrecognized application of the axiom of causality is in the Mori–Zwanzig projection operator formalism – see Zwanzig (2001, Chapter 8). This formalism leads in the linear response limit, to an exact reformulation of the response of a system to time-dependent dissipative fields in the form of a frequency- and wave vector-dependent generalized Langevin equation. In the time domain, the memory kernel associated with the generalized friction coefficient is convolved *backward* in time with the time-dependent driving force. This breaks the time reversal symmetry inherent in the equations of motion themselves. The temporal convolution is over the half space that describes history rather than the future. The spatial convolution, on the other hand, is over *all* physical space: $\pm\infty$ in each Cartesian dimension.

The axiom of causality is also met in electrodynamics where Maxwell's equations permit two solutions for the vector potential: the advanced and the retarded vector potential. In a well-known textbook, they state with little fanfare: "We can now neglect the term V_2' … for it would make the effect appear before the cause" Corson and Lorrain (1962, p. 445). Panofsky and Phillips (1969) are a little more equivocal on the subject: "but only the minus sign appears to have physical significance"; "the advanced potential … appears to violate elementary notions of causality."

It is interesting to re-examine the Boltzmann equation in the light of these observations. In writing the collision integral in the Boltzmann equation, it is assumed that, *before* collisions of ideal gas atoms, the positions and momenta are uncorrelated. After the collision there is correlation. The collision causes the *post*-collisional correlation. The cause of correlation is the collision, which occurs *before* the effect, which is correlation. In a causal Universe, the cause precedes the effect. This is consistent with the assumption of molecular chaos: *stosszahlansatz*.

If one assumes that the positions and momenta are correlated *before* the collisions, then one forms an anti-Boltzmann equation. This is exactly what one would expect if the Universe was, in fact, anti-causal where the coordinates and momenta *before* the collision are affected by the *later* collision. The effect *precedes* the cause, which is the collision.

So in an anti-causal Universe, dilute gases would be described by this anti-Boltzmann equation and the signs of all the transport coefficients (e.g., shear viscosity or thermal conductivity, etc.) would be opposite to those predicted from the Boltzmann equation. This reversal of signs of the transport coefficients for the anti-Boltzmann equation was first pointed out by Cohen and Berlin (1960). The connection between causality and *stosszahlansatz* is new.

Finally, we argue that in an anti-causal Universe where the future influences the present, the inevitable presence of innately random quantum processes in the future, or indeed the exercise of free will in the future by intelligent beings, makes the present state of the Universe undefined. We argue that the only possible Universe where time increases is, in fact, causal. If time were to decrease rather than increase, an anti-causal Universe would appear identical to our own. So ultimately we live in the only possible Universe and the causal second "Law" behavior is, on average, the only physically possible behavior.

References

Bertrand, J.L.F. (1887) *Thermodynamique*, Gauthier-Villars, Paris.

Boltzmann, L. (1872) Further studies on thermal equilibrium among Gas molecules. *Akad. Wissen. Wien*, **66**, 275.

Buckingham, E. (1905) On certain difficulties which are encountered in the study of thermodynamics. *Philos. Mag.*, **9**, 208.

Carberry, D.M., Williams, S.R., Wang, G.M., Sevick, E.M., and Evans, D.J. (2004) The Kawasaki identity and the fluctuation theorem. *J. Chem. Phys.*, **121**, 8179–8182.

Clausius, R. (1865) Ueber verschiedene Für Die anwendungen bequeme formen Der hauptgleichungen Der mechanischen wärmtheorie. *Ann. Phys. Chem.*, **125**, 353.

Clausius, R. (1872) A contribution to the history of the mechanical theory of heat. *Philos. Mag. J. Sci.*, **43**, 106–115.

Cohen, E.G.D. and Berlin, T.H. (1960) Note on the derivation of the Boltzmann equation from the Liouville equation. *Physica*, **26**, 717.

Corson, D.R. and Lorrain, P. (1962) *Introduction to Electromagnetic Fields and Waves*, W. H. Freeman and Company, San Francisco, CA.

Crooks, G.E. (1998) Nonequilibrium measurements of free energy differences for microscopically reversible Markovian systems. *J. Stat. Phys.*, **90**, 1481–1487.

Ehrenfest, P. and Ehrenfest, T. (1990) *The Conceptual Foundations of the Statistical*

Approach to Statistical Mechanics, Dover, Mineola, NY.

Evans, D.J., Cohen, E.G.D., and Morriss, G.P. (1993) Probability of 2nd Law violations in shearing steady-states. *Phys. Rev. Lett.*, **71**, 2401–2404.

Evans, D.J. and Morriss, G.P. (1990) *Statistical Mechanics of Nonequilibrium Liquids*, Academic Press, London.

Evans, D.J. and Searles, D.J. (1994) Equilibrium microstates which generate second Law violating steady states. *Phys. Rev. E*, **50**, 1645–1648.

Evans, D.J. and Searles, D.J. (1995) Steady states, invariant measures, and response theory. *Phys. Rev. E*, **52**, 5839–5848.

Evans, D.J. and Searles, D.J. (1996) Causality, response theory, and the second Law of thermodynamics. *Phys. Rev. E*, **53**, 5808–5815.

Evans, D.J., Searles, D.J., and Williams, S.R. (2008a) On the fluctuation theorem for the dissipation function and its connection with response theory. *J. Chem. Phys.*, **128**, 014504.

Evans, D.J., Searles, D.J., and Williams, S.R. (2008b) On the fluctuation theorem for the dissipation function and its connection with response theory [*J. Chem. Phys.* (2008) **128** 014504], Erratum.. *J. Chem. Phys.*, **128**, 249901.

Evans, D.J., Searles, D.J., and Williams, S.R. (2009a) Dissipation and the relaxation to equilibrium. *J. Stat. Mech: Theory Exp.*, P07029/1–P07029/11.

Evans, D.J., Searles, D.J., and Williams, S.R. (2009b) A simple mathematical proof of Boltzmann's equal a priori probability hypothesis, in *Diffusion Fundamentals III* (eds C. Chmelik, N. Kanellopoulos, J. Karger, and T. Doros), Leipziger Universitatsverlag, Leipzig.

Evans, D.J., Searles, D.J., and Williams, S.R. (2010) On the probability of violations of Fourier's Law for heat flow in small systems observed for short times. *J. Chem. Phys.*, **132**, 024501-1.

Evans, D.J., Williams, S.R., and Rondoni, L. (2012) A mathematical proof of the zeroth "Law" of thermodynamics and the nonlinear fourier "law" for heat flow. *J. Chem. Phys.*, **137**, 194109.

Evans, D.J., Williams, S.R., and Searles, D.J. (2011) A proof of Clausius' theorem for time reversible deterministic microscopic dynamics. *J. Chem. Phys.*, **134**, 204113-1.

Gibbs, J.W. (1981) *Elementary Principles in Statistical Mechanics*, Ox Bow Press, Woodbridge, CT.

Jarzynski, C. (1997) Nonequilibrium equality for free energy differences. *Phys. Rev. Lett.*, **78**, 2690–2693.

Lebowitz, J.L. (1993) Macroscopic laws, microscopic dynamics, Time's arrow and Boltzmann's entropy. *Physica A*, **194**, 1.

Loschmidt, J. (1876) Über Den zustand Des wärmegleichgewichtes eines systems Von kőrpern Mit rücksicht Auf Die schwerkraft I. *Sitzungsber. Kais. Akad. Wien. Math. Naturwiss. II*, **73**, 128–142.

Morriss, G.P. and Evans, D.J. (1985) Isothermal response theory. *Mol. Phys.*, **54**, 629–636.

Orr, W.M. (1904) On Clausius's theorem for irreversible cycles, and on the increase of entropy. *Philos. Mag. Ser.*, **6** (8), 509.

Orr, W.M. (1905) On Clausius' theorem for irreversible cycles, and on the increase of entropy. *Philos. Mag.*, **9**, 728.

Panofsky, W.K.H. and Phillips, M. (1969) *Classical Electricity and Magnetism*, Addison-Wesley, Reading, MA.

Petersen, C.F., Evans, D.J., and Williams, S.R. (2013) The instantaneous fluctuation theorem. *J. Chem. Phys.*, **139**, 184106.

Planck, M. (1905) On Clausius' theorem for irreversible cycles, and on the increase of entropy. *Philos. Mag.*, **9**, 167.

Searles, D.J. and Evans, D.J. (2000a) Ensemble dependence of the transient fluctuation theorem. *J. Chem. Phys.*, **113**, 3503–3509.

Searles, D.J. and Evans, D.J. (2000b) The fluctuation theorem and green-Kubo relations. *J. Chem. Phys.*, **112**, 9727–9735.

Searles, D.J. and Evans, D.J. (2004) Fluctuations relations for nonequilibrium systems. *Aust. J. Chem.*, **57**, 1119–1123.

Searles, D.J., Rondoni, L., and Evans, D.J. (2007) The steady state fluctuation relation for the dissipation function. *J. Stat. Phys.*, **128**, 1337–1363.

Sinai, Y.G. (1976) *Introduction to Ergodic Theory*, Princeton University Press, Princeton, NJ.

Thomson, W. (1874) Kinetic theory of the dissipation of energy. *Nature*, **9**, 441.

Wang, G.M., Sevick, E.M., Mittag, E., Searles, D.J., and Evans, D.J. (2002) Experimental demonstration of violations of the second Law of thermodynamics for small systems and short time scales. *Phys. Rev. Lett.*, **89**, 050601.

Williams, S.R., Searles, D.J., and Evans, D.J. (2006) Numerical study of the steady state fluctuation relations Far from equilibrium. *J. Chem. Phys.*, **124**, 194102-1.

Zwanzig, R. (2001) *Nonequilibrium Statistical Mechanics*, Oxford University Press, Oxford.

2
Introduction to Time-Reversible, Thermostatted Dynamical Systems, and Statistical Mechanical Ensembles

I have found it convenient, instead of considering one system of material particles, to consider a large number of systems similar to each other in all respects, except the initial circumstances of the motion, which are supposed to vary from system to system, the total energy being the same in all.

(Maxwell, 1879)

2.1
Time Reversibility in Dynamical Systems

Consider an isolated Hamiltonian system of interacting particles. The microscopic state of the system is represented by a phase space vector of the coordinates and canonical momenta of all the particles, in an exceedingly high dimensional space – phase space – $\{\mathbf{q}_1, \mathbf{q}_2, \ldots, \mathbf{q}_N, \mathbf{p}_1, \ldots, \mathbf{p}_N\} \equiv (\mathbf{q}, \mathbf{p}) \equiv \mathbf{\Gamma}$, where $\mathbf{q}_i, \mathbf{p}_i$ are the position and conjugate momentum of the particle i. The equations of motion for the system with an autonomous Hamiltonian $H(\mathbf{q}, \mathbf{p})$ are

$$\dot{\mathbf{q}}_i = \frac{\partial H(\mathbf{q}, \mathbf{p})}{\partial \mathbf{p}_i}$$
$$\dot{\mathbf{p}}_i = -\frac{\partial H(\mathbf{q}, \mathbf{p})}{\partial \mathbf{q}_i} \tag{2.1}$$

Definition

We define the *time reversal mapping* M^T, ..., as

$$[M^T\mathbf{\Gamma}] \equiv \mathbf{\Gamma}^T \equiv [M^T(\mathbf{q}, \mathbf{p})] \equiv (\mathbf{q}, -\mathbf{p}) \tag{2.2}$$

where the square brackets denote the fact that the time reversal operator, in this case, only attacks the phase space vector to the right. In general, operators attack anything that appears to their right (i.e., $M^T\mathbf{\Gamma} = \mathbf{\Gamma}^T M^T$).

The Hamiltonian for a system of interacting point particles, $H(\mathbf{q}, \mathbf{p}) = \sum_i p_i^2/2m + \Phi(\mathbf{q})$, is even in their momenta. We see from the equations of motion

Fundamentals of Classical Statistical Thermodynamics: Dissipation, Relaxation and Fluctuation Theorems, First Edition. Denis J. Evans, Debra J. Searles, and Stephen R. Williams.

that

$$\dot{\mathbf{\Gamma}} \equiv iL(\mathbf{\Gamma})\mathbf{\Gamma} = \left(\dot{\mathbf{\Gamma}} \cdot \frac{\partial}{\partial \mathbf{\Gamma}}\right)\mathbf{\Gamma} = \left(\frac{\mathbf{p}}{m} \cdot \frac{\partial}{\partial \mathbf{q}} - \frac{\partial \Phi}{\partial \mathbf{q}} \cdot \frac{\partial}{\partial \mathbf{p}}\right)(\mathbf{q}, \mathbf{p}) = (\mathbf{p}/m, -\partial \Phi / \partial \mathbf{q}) \tag{2.3}$$

where $iL \ldots \equiv \dot{\mathbf{\Gamma}} \cdot \partial \ldots / \partial \mathbf{\Gamma}$ is the time derivative operator for phase functions, and $\dot{\mathbf{\Gamma}}$ is given by the equations of motion (e.g., Eq. (2.1)).

Definition

We refer to $iL(\mathbf{\Gamma})$ as the *p-Liouvillean* or *phase-Liouvillean.*

The formal solution of Eq. (2.3) is

$$S^t\mathbf{\Gamma} \equiv \exp[iL(\mathbf{\Gamma})t]\mathbf{\Gamma} \tag{2.4}$$

and

$$\begin{aligned} \frac{d}{dt}(S^t\mathbf{\Gamma}) &= iL(\mathbf{\Gamma}) \exp[iL(\mathbf{\Gamma})t]\mathbf{\Gamma} = iL(\mathbf{\Gamma})S^t\mathbf{\Gamma} \\ &= \exp[iL(\mathbf{\Gamma})t]iL(\mathbf{\Gamma})\mathbf{\Gamma} = S^t\dot{\mathbf{\Gamma}} \end{aligned} \tag{2.5}$$

Definition

We refer to $\exp[iL(\mathbf{\Gamma})t]$ as the *p-propagator* or the *phase space propagator.*

If we apply the time reversal map to the *p*-Liouvillean, we see that

$$\begin{aligned} M^T iL(\mathbf{\Gamma}) \cdots &\equiv M^T \left(\dot{\mathbf{\Gamma}} \cdot \frac{\partial}{\partial \mathbf{\Gamma}}\right) \cdots = M^T \left(\frac{\mathbf{p}}{m} \cdot \frac{\partial}{\partial \mathbf{q}} - \frac{\partial \Phi}{\partial \mathbf{q}} \cdot \frac{\partial}{\partial \mathbf{p}}\right) \cdots \\ &= \left(-\frac{\mathbf{p}}{m} \cdot \frac{\partial}{\partial \mathbf{q}} + \frac{\partial \Phi}{\partial \mathbf{q}} \cdot \frac{\partial}{\partial \mathbf{p}}\right) M^T \cdots \\ &= -iL(\mathbf{\Gamma})M^T \cdots \end{aligned} \tag{2.6}$$

Using this result, we can apply the time reversal map to a propagated phase:

$$M^T(\exp[iL(\mathbf{\Gamma})t]\mathbf{\Gamma}) = \exp[-iL(\mathbf{\Gamma})t]\mathbf{\Gamma}^T \tag{2.7}$$

Definition

Time reversible dynamics satisfies the equation

$$M^T \exp(iL(\mathbf{\Gamma})t)M^T \exp(iL(\mathbf{\Gamma})t)\mathbf{\Gamma} = \mathbf{\Gamma} \tag{2.8}$$

This is proved by the substitution of Eq. (2.7):

$$\begin{aligned} &M^T \exp(iL(\mathbf{\Gamma})t)M^T \exp(iL(\mathbf{\Gamma})t)\mathbf{\Gamma} \\ &\quad = M^T \exp(iL(\mathbf{\Gamma})t) \exp(-iL(\mathbf{\Gamma})t)\mathbf{\Gamma}^T \\ &\quad = M^T\mathbf{\Gamma}^T = \mathbf{\Gamma} \end{aligned} \tag{2.9}$$

We will say in words what time reversibility entails. We start at a point in phase space, and evolve that phase forward in time by an amount t; reverse the signs of all the momenta, leaving the coordinates fixed; go forward in time using the same equations of motion for a duration t, and finally reverse all the momenta once again; then we end up at the point in phase space where we originally started.

We will add a few remarks about the notation. The S^t notation in Eqs. (2.4) and (2.5) for the propagator hides much subtlety. It is understood that it represents the p-propagator for the phase vector to its immediate right. In more complicated problems, such as those discussed in Eq. (2.9), the notation is ambiguous and inadequate. Is S^t the p-propagator for $\mathbf{\Gamma}$ or for $M^T\mathbf{\Gamma}$? We will use it only for the simplest of problems where there is no ambiguity. In these cases, the notation is very compact.

If we take the solar system and reverse all the momenta and angular momenta of the planets, then the resulting dynamics is also a solution of the equations of motion. If you watch a movie of the planets going around the sun and then play that movie backward, the resulting motion is still a solution of Hamilton's equations of motion and would not look un-physical.

However, if we do the same to the motion of a waterfall or a jet aircraft taking off, although the time-reversed movie is still a solution of the dynamical equations of motion, the time-reversed movie of a waterfall violates the second "Law" of thermodynamics. The time-reversed movie of the jet plane would constitute a perpetual motion machine of the second kind, thereby violating the second "Law" of thermodynamics. This is the so-called irreversibility paradox first pointed out by Kelvin and very shortly afterward by Loschmidt. The resolution of this paradox forms one of the main themes of this book.

If we watch the time-reversed movie of the solar system, then while observing the time-reversed planetary orbits (at least at a planetary length scale!) we would not see anything that would violate the second "Law" of thermodynamics. Thus the second "Law" of thermodynamics is somehow coupled to the complexity of the system. Maxwell was the first to realize this point (see the quote at the beginning of Chapter 3). As we will see, the fluctuation theorem proved in Chapter 3 will resolve these apparent paradoxes. In the process, the fluctuation theorem obviates the need for the second "Law" of thermodynamics.

2.2 Introduction to Time-Reversible, Thermostatted Dynamical Systems

Most systems of thermodynamic interest are not isolated. The vast majority of engines, devices, and all biological organelles exchange heat back and forth with their surroundings. These surroundings can often be viewed as being vastly larger than the engine, or the system of interest.

Think of the operation of an automobile engine. Ultimately, the chemical energy in the fuel is on average dissipated as heat to the surrounding atmosphere and earth. In principle, the size of the surroundings can be expanded virtually without limit. So, sufficiently far away from the system of interest, we can regard the surroundings as being unperturbed by the system of interest and, as we shall see, these unchanging surroundings can be regarded as being in a state of unchanging thermodynamic equilibrium. Later, we will learn how to treat cases where the surroundings are of the same size as the system of interest and therefore may also be out of equilibrium; but for the moment we will consider a nonequilibrium system of interest in contact with a much larger equilibrium reservoir.

For the moment, we regard an equilibrium system as a quiescent stationary state where, on average, no work is performed on the system and, on average, no net heat flows into or out of the system. Later (Section 4.2 and Chapter 5), we will be able to give a formal mathematical definition of an equilibrium system.

Because in Chapter 5 we will derive equilibrium relaxation theorems and through them the form of the equilibrium distributions of the microstates for an N-particle system, we will not employ any knowledge of that equilibrium distribution prior to Chapter 5.

We assume that classical mechanics gives an adequate description of the dynamics. We assume quantum and relativistic effects can be ignored. We also assume that the total momentum of the system of interest and, separately, the surroundings is zero. These systems are not in motion relative to each other or to the observer.

Experimentally we can only control a small number of variables that specify the *macroscopic* state of the system. We might only be able to control the system energy or the average kinetic energy of all or some of the particles, the volume V, and the number of atoms N in the system, which we assume to be constant. There is therefore an enormous range of microstates that are consistent with the small number of macroscopic constraints.

In writing the microscopic equations of motion for the system, it will be convenient to decompose the total system into two subsystems: the system of interest, and the surroundings. As we have pointed out earlier, the surroundings may often be regarded as being unperturbed by the system of interest. Conversely, provided the surroundings are not moving with respect to either the observer or the system of interest, and provided they have an unchanging distribution of microstates, the precise details of the microscopic equations of motion, or indeed the nature of the particles that constitute those surrounding systems, have no impact on the system of interest. The surrounding particles are too far from the system of interest.

The operation of an automobile is unaffected by the *microscopic details* of the road and atmosphere. Only a few macroscopic properties are important: average chemical composition, temperature and pressure of the air, and so on.

A typical experiment of interest is conveniently summarized by the following example. Consider an electrical conductor (e.g., a molten salt) subjected, at say $t = 0$, to an applied electric field $\mathbf{E}$. We wish to understand the behavior of this system from an atomic or molecular point of view. As in a laboratory, the molten salt is contained in a solid, electrically insulating conduction cell, and this cell is allowed to exchange heat with the much larger surroundings, so that once the system has reached a steady state, the average kinetic energy of the particles in the system of interest is constant.

If we use purely Hamiltonian equations of motion, the entire system will eventually heat up. We will need to supplement the Hamiltonian equations of motion with some time-reversible non-Hamiltonian terms buried deep in the surroundings so that a true nonequilibrium steady state is possible. The work that is done on the system is, on average, converted into heat, which is conducted through the system of interest and the walls, eventually getting removed on average by these non-Hamiltonian terms in the remote boundaries. Because these non-Hamiltonian terms are physically remote from the system of interest, there is no way that the system of interest can "know" how the heat is eventually removed.

The first time-reversible, deterministic thermostats and ergostats were invented simultaneously but independently by Hoover, Ladd, and Moran (1982), and Evans (1983) in the early 1980s. Prior to this development, there was no satisfactory mathematical way of modeling thermostatted nonequilibrium steady states.

Definition

We could study the macroscopic behavior of the macroscopic system by taking just one of the huge number of microstates that satisfy the macroscopic conditions and then solving the equations of motion for this single microscopic trajectory. We could then compute *time averages* $\overline{A}_t$ of a phase function $A(\mathbf{\Gamma})$:

$$\overline{A}_t \equiv \frac{1}{t}\int_0^t dsA(S^s\mathbf{\Gamma}) \tag{2.10}$$

However, we would have to take care that our microscopic trajectory $S^t\mathbf{\Gamma}$ was a *typical* trajectory, and that it did not behave in an exceptional way.

Definition

Perhaps a better way of understanding the macroscopic system would be to select a set of N_Γ initial phases (i.e., microstates) $\{\mathbf{\Gamma}_j, j = 1, \dots, N_\Gamma\}$ distributed according to the naturally occurring states that are consistent with the small number of macro-constraints and compute the time-dependent properties of the macroscopic system by taking a time-dependent *ensemble average* $\langle A(t)\rangle$ of a phase function $A(\mathbf{\Gamma})$ over the *ensemble* of time-evolved phases $\langle A(t)\rangle = \lim_{N_\Gamma\to\infty} \sum_{j=1}^{N_\Gamma} A(S^t\mathbf{\Gamma}_j)/N_\Gamma$.

Indeed, repeating the experiment with initial states that are consistent with the specified initial conditions is often what an experimentalist does in the laboratory (to within limits of experimental capacity to control the initial conditions). One could then try to specify the initial phase space probability density consistent with the known, small number of macroscopic conditions $f(\mathbf{\Gamma}; 0)$, and try to understand the time-dependent evolution of this density $f(\mathbf{\Gamma}; t)$. Time-dependent averages of phase functions could then be computed as a time-dependent ensemble average:

$$\langle A(t)\rangle = \int d\mathbf{\Gamma}A(\mathbf{\Gamma})f(\mathbf{\Gamma}; t) = \int d\mathbf{\Gamma}A(S^t\mathbf{\Gamma})f(\mathbf{\Gamma}; 0) \tag{2.11}$$

The equality of the average over the initial distribution with that taken over the time-dependent distribution is guaranteed by the conservation of probability. Although the concept of ensemble averaging seems natural and intuitive to experimental scientists, the use of ensembles has caused some problems and misunderstandings from a more purely mathematical viewpoint.

Definition

A system is said to be *time-stationary* or simply *stationary* if the ensemble averages appearing in Eq. (2.11) are independent of time: $\langle A(t)\rangle = \langle A(t+\Delta)\rangle,\ \ \forall\Delta > 0$.

Definition

A stationary system is said to be *physically ergodic* if the time average of a phase function representing a physical observable, along a trajectory that starts almost anywhere in the ostensible phase space, $\forall!\boldsymbol{\Gamma} \in D$, is equal to the ensemble average taken over an ensemble of systems consistent with the small number of macroscopic constraints on the system.

$$\lim_{t\to\infty}\langle A(t)\rangle = \lim_{t\to\infty}\overline{A}_t(\boldsymbol{\Gamma}) = \lim_{t\to\infty}\frac{1}{t}\int_0^t ds\ A(S^s\boldsymbol{\Gamma}),\quad \forall!\boldsymbol{\Gamma}\in D \tag{2.12}$$

The system may be stationary for all values of time, or it may only be asymptotically stationary in the limit of long time – as implied by our notation. In this text, this is what is meant when a system is referred to as *ergodic*.

Experience shows that, for an isolated Hamiltonian system of interacting particles with no applied dissipative fields, the system will usually relax to a time-stationary state where time averages of almost all macroscopic variables such as pressure or density become time-independent. That state is called the *state of microcanonical equilibrium*. Similarly, if a Hamiltonian system free of applied dissipative fields (such as the electric field for electrically conductive systems) is allowed to exchange heat with a vastly larger heat bath which itself can be considered to be at equilibrium, then at long times the Hamiltonian system will be expected to relax to the canonical equilibrium state. Later, in Chapter 5 of this book, we will (subject to some fairly simply stated mathematical conditions) prove equilibrium relaxation theorems, which show that initial nonequilibrium systems will at long times relax, perhaps nonmonotonically, toward a physically ergodic state of thermodynamic equilibrium. Those same theorems also give precise mathematical expressions for the equilibrium phase space distributions, both canonical and microcanonical.

Ensembles are well known in equilibrium statistical mechanics, the concept being first introduced by Boltzmann (1872) and later by Maxwell (1879). The use of ensembles in nonequilibrium statistical mechanics is less widely known and understood.[1] In our analyses, it will often be convenient to choose the initial ensemble that is represented by the set of phases $\{\boldsymbol{\Gamma}_j, j = 1, \ldots, N_\Gamma\}$ to be one of the standard ensembles of equilibrium statistical mechanics. However, sometimes we may wish to vary this somewhat. In any case, in all the examples we will consider,

1) For further background information on nonequilibrium statistical mechanics, see Evans and Morriss (1990).

the initial ensemble of phase vectors will be characterized by a *known* initial N-particle distribution function $f(\mathbf{\Gamma};0)$, which gives the probability $f(\mathbf{\Gamma};0)d\mathbf{\Gamma}$ that a member of the ensemble is within some infinitesimal neighborhood $d\mathbf{\Gamma}$ of a phase $\mathbf{\Gamma}$ at time 0. By construction, the number of ensemble members is conserved.

Consider an electric field, which, on average, does work on an electrically conducting system, causing an electric current $\mathbf{I} \equiv \sum c_i \dot{\mathbf{q}}_i$ to flow (c_i is the electric charge on particle i). To remove the complicating effects of local charge buildup or surface electrolysis, we employ periodic boundary conditions in the direction of the electric field. This will allow the current to flow forever and also allow for the possibility of establishing a nonequilibrium time-stationary state, or a nonequilibrium steady state.

It is exceedingly important to remember that we are *expressly* excluding the case where the system is an insulator and the field induces a polarization rather than a current! The difference between an insulator and a conductor can be determined only by the physics of the situation. If we subject solid sodium chloride to an electric field at room temperature, then the field induces a polarization that explicitly changes the internal energy of the system (the expression for the internal energy includes a term that is function of the field). Electrostatic potential energy is stored in the system. It acts as a capacitor.

If we make the single change of increasing the temperature of sodium chloride to 1100 K, then sodium chloride melts and becomes an electrical conductor. The electric field does not explicitly change the internal energy of the system. So the difference between an insulator and a conductor cannot be determined from the equations of motion! The difference can be in the initial conditions for the equations of motion – in this case the initial energy or temperature. We will treat the case where fields change the internal energy of the system in Chapter 8. Until then, it is assumed that the external fields are purely dissipative and do not explicitly change the internal energy of systems.

Definition

If external fields are applied to the system of particles and the external field does work on the system and if that work can be turned *completely* into heat that can diffuse out of the system, the external field is termed a *dissipative field.* If the work can be completely stored in the system in the form of potential energy, the external field is termed *nondissipative*.

The electric field applied to solid sodium chloride is nondissipative because application of a field results in energy being stored in the system. Another example of a nondissipative field is application of strain to an *elastic* solid. The work is stored in the intermolecular potential energy of the solid's constituent molecules.

Considering the system of charged particles again, from experience we expect that at an arbitrary time t after a dissipative field has been applied, the ensemble-averaged electric current $\langle \mathbf{I}(t) \rangle$ will be in the direction of the field; the work performed on the system by the field will be transformed into (or dissipated as) heat – Ohmic heating, $\langle \mathbf{I}(t) \rangle \cdot \mathbf{E}$. It will frequently be the case that the electrical conductor will be in contact with an electrically insulating heat reservoir that fixes the average energy of the system so that, on average, heat flows from the system of interest, the conduction cell, toward the much larger heat reservoir. Nonetheless, all the particles in this system (conduction cell plus reservoir) constitute a time-reversible dynamical system. The equations of motion, whether quantum or classical, are time-reversal-symmetric.

We are interested in a number of problems suggested by this experiment:

1) How do we reconcile the Ohmic heating with the time reversibility of the microscopic equations of motion? Why is there no possibility of Ohmic cooling?
2) For a given initial phase $\mathbf{\Gamma}_j$ that generates some time-dependent current $\mathbf{I}(S^t\mathbf{\Gamma}_j)$, can we generate Loschmidt's conjugate *antitrajectory*, which has a time-reversed electric current with reversed time ordering?
3) Is there anything we can say about the deviations of the behavior of individual ensemble members from the average behavior?

In noting these three questions, question 2 is slightly different from what is usually mentioned in textbooks that discuss reversibility. For the antitrajectory, not only is the current of the opposite sign to that for its conjugate trajectory but also the time-ordered fluctuations and transients must exhibit reversed time ordering. The last temporal fluctuations that occur on a particular trajectory are, in fact, the first fluctuations on its conjugate antitrajectory. This, as we will see in Chapter 9, is hugely important.

Consider a classical system of N interacting particles in a three-dimensional volume V. Initially (at $t=0$), the microstates of the system are distributed according to a given normalized probability distribution function $f(\mathbf{\Gamma};0)$. To apply our results to realistic systems, we separate the N-particle system into a system of interest and a wall thermostatting region containing N_{th} particles. Note: the total number of wall particles is N_W and the system of interest may contain unthermostatted wall particles. Within the thermostat, particles are subject to a fictitious thermostat or an ergostat. The thermostat employs a switch, $S_i = 1,\ 0$, which controls how many and which particles are thermostatted: $S_i = 0$ if $1 \le i \le (N - N_{\text{th}})$; $S_i = 1$ if $(N - N_{\text{th}} + 1) \le i \le N$; and $N_{\text{th}} \le N_W$. We define the thermostat kinetic energy

$$K_{\text{th}} \equiv \sum_{i=1}^{N} S_i \frac{p_i^2}{2m_i} \tag{2.13}$$

and write the equations of motion for the composite N-particle system as

$$\begin{aligned}
\dot{\mathbf{q}}_i &= \frac{\mathbf{p}_i}{m_i} + \mathbf{C}_i(\mathbf{\Gamma}) \cdot \mathbf{F}_e \\
\dot{\mathbf{p}}_i &= \mathbf{F}_i(\mathbf{q}) + \mathbf{D}_i(\mathbf{\Gamma}) \cdot \mathbf{F}_e - S_i(\alpha \mathbf{p}_i + \mathbf{F}_{\text{th}}) \\
\dot{\alpha} &= \left[\frac{2K_{\text{th}}}{3N_{\text{th}} k_B T_{\text{th}}} - 1 \right] \frac{1}{\tau^2}
\end{aligned} \tag{2.14}$$

where $\mathbf{F}_i(\mathbf{q}) = -\partial\Phi(\mathbf{q})/\partial\mathbf{q}_i$ is the interatomic force on particle i, $\Phi(\mathbf{q})$ is the interparticle potential energy, and $\mathbf{C}_i(\mathbf{\Gamma})$ and $\mathbf{D}_i(\mathbf{\Gamma})$ are tensorial phase functions that couple the dissipative field to the system of interest. The other terms are explained in the following paragraphs.

It needs to be noted that some external fields are odd under the time-reversal map (e.g., the strain rate) while others are even (e.g., an electric field). In some experimental or computer simulation circumstances, it may be inconvenient to invert the sign of a dissipative field that is odd under M^T. In such cases, other mappings such as the Kawasaki map for shear flow – see Section 7.4 of Evans and Morriss (1990) – may be used in place of the time-reversal map in order to construct the conjugate antitrajectories with the sign of the dissipative field being unchanged. Any means of generating the conjugate antitrajectories may be used in place of the standard time-reversal map.

The term involving $-S_i\alpha\mathbf{p}_i$ is a deterministic, time-reversible Nosé–Hoover thermostat (Evans, 1985; Hoover, 1985), which is used to add or remove heat from the particles in the reservoir region through introduction of an extra degree of freedom described by α; T_{th} is a *target* parameter that controls the time-averaged kinetic energy of the thermostatted particles; and τ is the time constant for the integral feedback mechanism of the Nosé–Hoover thermostat. If $2K_{\text{th}} > 3N_{\text{th}}k_BT_{\text{th}}$, then $\dot{\alpha} > 0$ and the thermostat multiplier will increase, implying that in the future more energy will be removed from the thermostatted particles. Conversely, if $2K_{\text{th}} < 3N_{\text{th}}k_BT_{\text{th}}$, then $\dot{\alpha} < 0$, implying that less energy will be removed by the thermostat in the future. Thus the thermostat tends to stabilize the average thermostat kinetic energy at the value $\overline{K}_{\text{th}} = 3N_{\text{th}}k_BT_{\text{th}}/2$.

It is a trivial matter to check that the Nosé–Hoover thermostatted equations of motion are time-reversal-symmetric. From the third equation in Eq. (2.14), we see that $\dot{\alpha}$ is even under time reversal. This means that α is odd, implying that the whole thermostatting term in the $\dot{\mathbf{p}}$ equation of motion is even as, of course, is the force.

Assuming that the system comes to a nonequilibrium (stationary) steady state where at long times time averages of smooth phase functions become time-independent, we also expect there will be a time-independent value for the thermostat multiplier $\lim_{t\to\infty}\langle\alpha(t)\rangle = \langle\alpha\rangle_\infty \Rightarrow \lim_{t\to\infty}\langle\dot{\alpha}(t)\rangle = 0$. From Eq. (2.14), we see that in this steady state

$$\lim_{t\to\infty}\frac{1}{t}\int_0^t ds\,K_{\text{th}}(s) \equiv \overline{K}_{\text{th}} = 3N_{\text{th}}k_BT_{\text{th}}/2 \tag{2.15}$$

In Chapter 5, we will prove that, if the dissipative field is in fact zero and the system is T-mixing, the system described by Eq. (2.14) eventually comes to thermodynamic equilibrium and the target temperature of the Nosé–Hoover thermostat, T_{th}, is then identical to the equilibrium thermodynamic temperature of the system. When the dissipative field is nonzero, the classically defined thermodynamic temperature of the system of interest is, in fact, *undefined.* However, in this case, if the thermal reservoirs are made arbitrarily large compared to the system of interest, the thermal reservoirs will be hardly affected by the system of interest and T_{th} can be regarded as the equilibrium thermodynamic temperature of the thermal reservoir. For an in-depth discussion of the Nosé-Hoover thermostat, see Section 5.2 of Evans and Morriss (1990).

To model a realistic system, we can choose $\mathbf{C}_i(\mathbf{\Gamma}) = \mathbf{0}$ and $\mathbf{D}_i(\mathbf{\Gamma}) = \mathbf{0}$ when $S_i = 1$. This means that the dissipative field cannot do work on the thermostatted particles. In Eq. (2.14), the fluctuating force $\mathbf{F}_{\text{th}} = (1/N_{\text{th}}) \sum_{i=1}^{N} S_i \mathbf{F}_i$, which ensures that the macroscopic momentum of the thermostatted particles is a constant of the motion, which we set to zero.

The Nosé–Hoover thermostat adds a degree of freedom to the phase space of the system, and the extended phase space vector is $\mathbf{\Gamma}^* \equiv (\mathbf{\Gamma}, \alpha)$. However, in order to simplify the notation, from here on we represent this implicitly using $\mathbf{\Gamma}$.

Definition

If in the absence of the thermostatting terms, the equations of motion preserve the phase space volume, $\Lambda^{\text{ad}} \equiv (\partial/\partial\mathbf{\Gamma}) \cdot \dot{\mathbf{\Gamma}}^{\text{ad}} = 0$, and the system is said to satisfy the condition known as the *adiabatic incompressibility of phase space* or AI$\mathbf{\Gamma}$, (Evans and Morriss, 1990). All systems studied in this book satisfy this condition.

If we set $S_i = 1$ for the particles in the surroundings (e.g., the walls) only, the equations of motion for the thermostatted particles are supplemented with unnatural thermostat and force terms, but the equations of motion for the particles in the system of interest are quite natural. Equations (2.14) are time-reversible and heat can be either absorbed or given out by the thermostat. Similar constructions have been applied in various studies. Of course, if $S_i = 1$ for all i, we obtain a homogeneously thermostatted system, which is often studied (Evans and Morriss, 1990).

The model system could be quite realistic with only some particles subject to the external field. For example, some particles might be charged in an electrical conduction experiment while others may be chemically distinct, being solid at the temperatures and densities under consideration. Furthermore, these particles may form the thermal boundaries or walls which thermostat and "contain" the electrically charged fluid particles inside a conduction cell. In this case, $S_i = 1$ only for wall particles and $S_i = 0$ for all the fluid particles. This would provide a realistic model of electrical conduction and is frequently used in nonequilibrium molecular dynamics (NEMD) computer simulations.

Definition

If we consider a group of atoms within some small volume δV centered on a position $\mathbf{r}$, the *local mass density* $\rho(\mathbf{r})$ is defined as (Evans and Morriss, 1990)

$$\rho(\mathbf{r}) \equiv \sum_{i \in \delta V} m_i/\delta V \tag{2.16}$$

and the *local streaming velocity* $\mathbf{u}$ is defined by the equation

$$\sum_{i \in \delta V} \mathbf{p}_i \equiv \rho(\mathbf{r})\mathbf{u}(\mathbf{r})\delta V \tag{2.17}$$

where $\mathbf{p}_i$ is the momentum of particle i, measured in the laboratory frame. Adjusting the physical size of the volume δV adjusts the special resolution within which we measure local properties.

It is worth pointing out that, if the streaming velocity varies significantly over the range of the interatomic forces and/or if the streaming velocity varies significantly over microscopic relaxation times, it becomes impossible to define and the hydrodynamic description given here simply breaks down. Examples include shock waves and extremely turbulent flows.

Definition

If the momenta and velocities are computed relative to the *local streaming velocity* (i.e., $m_i\dot{\mathbf{q}}_i - m_i\mathbf{u}(\mathbf{r}_i)$, $\dot{\mathbf{q}}_i - \mathbf{u}(\mathbf{r}_i)$), they are termed *peculiar momenta* and *peculiar velocities*, respectively.

The use of peculiar momenta in the expressions for the kinetic temperature and the internal energy is important. All thermodynamic variables must be independent of the velocity of the frame of reference from which they are measured. For instance, if we consider a glass of water in a moving train, the total energy of the molecules comprising the glass of water is dependent on the velocity of the train. However, the internal energy and the kinetic temperature are independent of the motion of the train. All thermodynamic quantities must be evaluated using momenta and velocities measured in the local streaming velocity frame of reference. In writing Eq. (2.14), we chose a frame of reference where the total momentum of the thermostat is initially zero, and define the fluctuating force $\mathbf{F}_{\text{th}}$ to make the momentum of the thermostat a constant of the motion. This means that the average momentum of the entire system is also zero, ensuring that all the momenta appearing in Eq. (2.14) are, in fact, peculiar.

Definition

For systems with no applied external field or for which the external field is purely dissipative, the Hamiltonian $H_0 \equiv \sum_{i:S_i=1} \left[p_i^2/2m + \Phi_i(\mathbf{q})\right]$ expressed in peculiar momenta $\mathbf{p}_i$ and where $\mathbf{\Phi}_i$ is the contribution to the interparticle energy from particle i, has an average value, which is the *thermodynamic internal energy* of the system. The internal energy is just the energy of the system with the local streaming kinetic energy removed. It is also important to note that, because the processes considered here are purely dissipative, the form of the Hamiltonian is independent of the value of the dissipative field. The internal energy has no *explicit* dependence on the external field. This is entirely analogous to autonomous Hamiltonians that contain no explicit reference to time. In Chapter 8, we will discuss cases where the field appears explicitly in the Hamiltonian.

This microscopic internal energy $H_0(\mathbf{\Gamma})$ is completely mechanical – as the so-called first "law" of thermodynamics shows. The definition of the internal energy is valid even far from equilibrium provided the streaming velocity Eq. (2.17) is well defined.

An alternative thermostatting mechanism is to choose the thermostat multiplier α to make either the internal energy of the thermostat

$H_0 \equiv \sum_{i:S_i=1} \left[p_i^2/2m + \Phi_i(\mathbf{q})\right]$, or of the entire system a constant of the motion. For this *ergostatted dynamics*, the thermostat multiplier α is chosen as the instantaneous solution to the equation

$$\begin{aligned}\dot{H}_0(\mathbf{\Gamma}) &\equiv -\mathbf{J}(\mathbf{\Gamma})V \cdot \mathbf{F}_e - 2K_{\text{th}}(\mathbf{\Gamma})\alpha(\mathbf{\Gamma}) \\ &\equiv -\mathbf{J}(\mathbf{\Gamma})V \cdot \mathbf{F}_e - \dot{Q}_{\text{th}} \\ &= 0 \end{aligned} \tag{2.18}$$

Definition

The heat *added to the thermostat* per unit time $\dot{Q}_{\text{th}}(t)$ is defined in Eq. (2.18).

[**Aside**
When the thermostat is overwhelmingly larger than the system of interest and when it is in thermodynamic equilibrium, the thermostat increases its entropy at a rate $\dot{S}_{\text{th}} = \dot{Q}_{\text{th}}/T_{\text{th}}$, where T_{th} is the equilibrium thermodynamic temperature of the large thermostat.]

A third thermostatting mechanism is where we make the peculiar kinetic energy of the thermostat

$$K_{\text{th}} \equiv \sum_{i=1}^{N_{\text{th}}} S_i p_i^2/2m = (3N_{\text{th}} - 4)k_B T_{\text{th}}/2 \tag{2.19}$$

a constant of the motion; in which case, we speak of Gaussian isokinetic dynamics. In Eq. (2.19), $N_{\text{th}} = \sum S_i$.

Definition

The quantity T_{th} defined by Eq. (2.19) is called the *kinetic temperature* of the thermostat.

Both these latter thermostatting methods involve differential feedback, and the equations of motion can be derived using Gauss' principle of least constraint to fix either the internal energy or the thermostat's peculiar kinetic energy (Evans *et al.*, 1983). In both cases, the first two equalities of Eq. (2.14) still apply but the third equality is replaced by an explicit expression for the multiplier. These Gaussian thermostats were, in fact, the first time-reversible deterministic thermostats that were proposed. Hoover *et al.* developed the isokinetic thermostat (Hoover, Ladd, and Moran, 1982), while Evans (simultaneously, in the same week) developed the ergostat (Evans, 1983). These thermostats were developed in order to construct NEMD computer simulation algorithms for the study of nonequilibrium systems and the calculation of transport coefficients (Evans and Morriss, 1990).

It is a trivial matter to check that these thermostats are time-reversal-symmetric. We note from Eq. (2.18) that, as in the Nosé–Hoover thermostat, α is odd under time reversal.

One might wonder whether other mathematical forms are possible for these Gaussian thermostats. Could one replace the thermostat term in the equation

$\dot{\mathbf{p}} = \mathbf{F} - \alpha\mathbf{p}$ with a so-called μ-thermostat, so $\dot{\mathbf{p}} = \mathbf{F} - \alpha|p|^{\mu-1}\mathbf{p}$, and choose α to fix either the kinetic energy, or following Gauss Principle, $\sum |p|^{\mu+1}$? It turns out that, if $\mu \neq 1$, these systems can never relax to equilibrium (Bright, Evans, and Searles, 2005; Evans, Searles, and Williams, 2010). If you consider the finite-difference form of the equations of motion for μ thermostats, you can see that only when $\mu = 1$ does the finite-difference form correspond to a rescaling of time so that in rescaled time the thermostat kinetic energy is fixed. When $\mu \neq 1$, this time rescaling profoundly alters the velocity distribution because the time rescaling factor is dependent of the speed of the different individual particles. Nosé–Hoover thermostats permit a greater range of choices for μ; however, we do not consider these $\mu \neq 1$ thermostats any further.

Definition

The *dissipative flux* $\mathbf{J}$ due to the driving *dissipative field* $\mathbf{F}_e$ is defined as (Evans and Morriss, 1990)

$$\dot{H}_0^{\text{ad}} \equiv -\mathbf{J}(\mathbf{\Gamma})V \cdot \mathbf{F}_e \equiv \sum \left[\frac{\mathbf{p}_i}{m} \cdot \mathbf{D}_i - \mathbf{F}_i \cdot \mathbf{C}_i\right] \cdot \mathbf{F}_e \tag{2.20}$$

$\dot{H}_0^{\text{ad}}$ is the adiabatic time derivative of the internal energy (i.e., it is computed without the contributions from the thermostat), and V is the volume of the system.

As mentioned earlier, it is always assumed that the equations of motion for the driven system satisfy the adiabatic incompressibility of phase space condition. In Chapter 6, we will show in detail why the dissipative flux is so named.

Definition

The *adiabatic time derivative* of H_0 ($\dot{H}_0^{\text{ad}}$) is, in fact, the *work* performed on the system by the dissipative field because it is the total change of energy minus the heat removed by the thermostat/ergostat.

Equation (2.18) is a statement of the first law of thermodynamics for an ergostatted nonequilibrium system. The energy removed from (or added to) the system by the ergostat must be balanced instantaneously by the work done on (or removed from) the system by the external dissipative field F_e. For ergostatted dynamics, we solve Eq. (2.18) for the ergostat multiplier and substitute this phase function into the equations of motion. For isokinetic dynamics, we solve an equation that is analogous to Eq. (2.18) but which ensures that the kinetic temperature of the walls or system is fixed. The equations of motion (Eq. 2.14) are reversible if the thermostat multiplier is defined in this way.

A simple example system is the case of electrical conductivity. There, we could model the charged ions of a molten salt ($\mathbf{C}_i = 0$, $\mathbf{D}_i = c_i\mathbf{I}$, where c_i is the electric charge of particle i and $\mathbf{I}$ is the identity matrix) subject to an electric field $\mathbf{F}_e = \mathbf{E}$. We could surround these ions with neutral atoms ($c_i = 0$) of a solid wall that contains the electrically conducting molten salt. Further, outside these realistically modeled wall particles, we could then have a layer of thermostatted or ergostatted, electrically neutral wall particles. These thermostatted particles can be located arbitrarily far from the system of interest.

One might object that our analysis is compromised by our use of these artificial (time-reversible) thermostats. Since, the thermostat can be made arbitrarily remote from the system of physical interest, the system cannot "know" the precise details of how the heat is ultimately removed (Evans, Searles, and Williams, 2009). This means that the results obtained for the system using our simple mathematical thermostat must be the same as the those we would infer for the same system surrounded (at a distance) by a real physical thermostat (say with a huge heat capacity). These mathematical thermostats may be unrealistic, but in the final analysis they are very convenient but ultimately irrelevant devices. Importantly, they allow us to do the mathematical bookkeeping that is necessary in the study of systems that exchange heat with their surroundings. Ultimately, the work that is, on average, performed on the system of interest is ultimately, on average, transformed into heat, which is absorbed by an infinitely large system that can be regarded as being arbitrarily close to thermodynamic equilibrium, and arbitrarily far from the system of interest. That reservoir has a known kinetic temperature, which, as we will see in Chapter 5, is the equilibrium thermodynamic temperature of that reservoir. Again, as will be proved in Chapter 5, this kinetic temperature is also the equilibrium temperature the entire system will relax to if it is allowed to do so.

2.3
Example: Homogeneously Thermostatted SLLOD Equations for Planar Couette Flow

A very important dynamical system is the standard model for thermostatted planar Couette flow – the so-called SLLOD equations for shear flow (Evans and Morriss, 1984a). Consider N particles under shear. In this system, the external field is the shear rate $\partial u_x/dy = \dot{\gamma}(t)$ (the y-gradient of the x-streaming velocity $u_x(y)$). The equations of motion for the particles are given by the so-called homogeneously thermostatted SLLOD equations

$$\dot{\mathbf{q}}_i = \mathbf{p}_i/m + \mathbf{i}\dot{\gamma}y_i, \quad \dot{\mathbf{p}}_i = \mathbf{F}_i - \mathbf{i}\dot{\gamma}p_{yi} - \alpha\mathbf{p}_i \tag{2.21}$$

Here, $\mathbf{i}$ is a unit vector in the positive x-direction. At low Reynolds numbers, where a planar velocity profile is expected to be stable, the SLLOD momenta are, in fact, peculiar momenta (i.e., they are measured relative to the average streaming velocity of the individual particles, $\mathbf{u}(\mathbf{q}_i, t) = \mathbf{i}\dot{\gamma}y_i$).

As first pointed out by Evans and Morriss (1984a), the adiabatic SLLOD equations of motion give an exact description of planar Couette flow arbitrarily far from equilibrium (Evans and Morriss, 1984a,b). This is because the adiabatic SLLOD equations for a step-function strain rate $\partial u_x(t)/\partial y = \dot{\gamma}(t) = \dot{\gamma}\Theta(t)$ are equivalent to Newton's equations after the impulsive imposition of a linear velocity gradient at $t=0$ (i.e., $d\mathbf{q}_i(0^+)/dt = d\mathbf{q}_i(0^-)/dt + \mathbf{i}\dot{\gamma}y_i(0)$) (Evans and Morriss, 1984b). There is thus a remarkable subtlety in the SLLOD equations of motion. If one starts at $t=0-$, with a canonical ensemble of systems then at $t=0+$, the SLLOD equations of motion transform this initial ensemble into the local equilibrium ensemble for planar Couette flow at a shear rate $\dot{\gamma}$ (Daivis and Todd, 2006).

Because the effects of thermostatting are asymptotically quadratic in the strain rate, the homogeneously thermostatted SLLOD equations of motion give an exact description of the *linear* response of a system to planar Couette flow – even a time-dependent planar Couette flow. For a mathematical proof, see Evans and Morriss (1990).

At low Reynolds numbers, the SLLOD momenta $\mathbf{p}_i$ are peculiar momenta, and α is determined using Gauss's principle of least constraint to keep the internal energy $H_0 = \sum p_i^2/2m + \Phi(\mathbf{q})$ fixed (Note: the internal energy is the sum of the peculiar kinetic energy and the potential energy. It is *not* the sum of the laboratory kinetic energy and the potential energy.). Thus, for our system Eq. (2.21)

$$\begin{aligned}\dot{H}_0 &= \sum \dot{\mathbf{p}}_i \cdot \frac{\partial K}{\partial \mathbf{p}_i} + \dot{\mathbf{q}}_i \cdot \frac{\partial \Phi}{\partial \mathbf{q}_i} \\ &= \sum \left(-\mathbf{i}\dot{\gamma} p_{yi} \cdot \frac{\mathbf{p}_i}{m} - \mathbf{i}\dot{\gamma} y_i \cdot \mathbf{F}_i\right) - 2K(\mathbf{p})\alpha \\ &= -\dot{\gamma} \sum \left(\frac{p_{yi} p_{xi}}{m} + y_i F_{xi}\right) - 2K(\mathbf{p})\alpha \equiv -\dot{\gamma} P_{xy}(\mathbf{q}, \mathbf{p}) V - 2K(\mathbf{p})\alpha \end{aligned} \tag{2.22}$$

where P_{xy} is a well-known microscopic expression for the xy element of the pressure tensor (Irving and Kirkwood, 1950) in a homogeneous system. We can fix the microscopic internal energy of the system by choosing α as

$$\alpha = -P_{xy}(\mathbf{q}, \mathbf{p})\dot{\gamma} V / 2K(\mathbf{p}) \tag{2.23}$$

From the equations of motion, we can then express the rate of change of internal energy for isoenergetic SLLOD dynamics in terms of the rate at which work is done on the system and the rate at which heat is extracted from the system:

$$\dot{H}_0(t) = -P_{xy}(t)\dot{\gamma} V - 2K\alpha(t) = -P_{xy}(t)\dot{\gamma} V - \dot{Q}(t) = 0 \tag{2.24}$$

Thus the ergostat increases the internal energy of the system at a rate $-\dot{Q}(t)$, which is precisely and instantaneously equal to the rate at which work is expended on the system by the shearing deformation, namely $-P_{xy}(t)\dot{\gamma} V$. Incidentally, this is precisely the viscous dissipation one finds for planar Couette flow from the Navier–Stokes equations.

The corresponding isokinetic form for the thermostat multiplier is obtained by setting $\dot{K} = 0$, which, using Eq. (2.21), gives

$$\alpha(t) = \frac{\sum_{i}^{N} \mathbf{F}_i \cdot \mathbf{p}_i - \dot{\gamma} \sum_{i=1}^{N} p_{xi} p_{yi}/m}{\sum_{i=1}^{N} \mathbf{p}_i^2/m} \tag{2.25}$$

We note that, in the Nosé–Hoover thermostatted forms for the SLLOD equations, $\dot{\alpha}$ is given, as usual, by Eq. (2.14).

If the thermostatted SLLOD system comes to a nonequilibrium steady state, the time-averaged rate of shearing work that is performed by the shear on the system is equal to the time averaged rate at which heat is removed by the thermostat from the system. If the system is ergostatted, this balance is achieved instantaneously.

Definition

The ergostatted and thermostatted SLLOD equations of motion (Eqs. (2.21)–2.23) are time-reversible (if $\dot{\gamma}$ is fixed, M^T should be replaced by the *Kawasaki mapping:* $M^K(\mathbf{q}_x, \mathbf{q}_y, \mathbf{p}_x, \mathbf{p}_y) = (\mathbf{q}_x, -\mathbf{q}_y, -\mathbf{p}_x, \mathbf{p}_y)$ in Eq. (2.8)).

In the weak flow limit, these equations yield the correct Green–Kubo relation for the linear shear viscosity of a fluid (Evans and Morriss, 1984b). We have also proved that in this limit the linear response obtained from the equations of motion, or equivalently from the Green–Kubo relation, are identical to leading order in N the number of particles.

In computer simulations, if one wants to carry out NEMD simulations one has to supplement the SLLOD equations of motion with appropriate boundary conditions – one cannot simulate infinite systems. If you start with a cubic periodic system at $t = 0$, then the shear motion causes the unit cells immediately above and below the primitive cell to slide to the right and left above and below the primitive cell at constant speeds $V^{1/3}\dot{\gamma}$. At time t, these cells move to positions $\pm V^{1/3}\dot{\gamma}t$. The position of these cells affects the forces $\mathbf{F}_i$ on the N particles in the primitive cell. This means that for finite periodic systems the SLLOD equations of motion as implemented are in fact non-autonomous!

In practice for the short range forces like Lennard-Jones or Weeks-Chandler-Andersen (WCA) (Weeks, Chandler and Andersen (1971)) forces these non-autonomous effects can be hard or nearly impossible to observe in three dimensional systems where $N \geq\sim 100$ (Petravic, 2005; Bernardi, Brookes, and Searles, 2014). In working with small systems, the nonautonomous effects can be easily observed particularly close to the freezing density.

2.4 Phase Continuity Equation

We have introduced the phase space distribution function $f(\mathbf{\Gamma}; t)$. It gives the probability per unit phase space volume of finding ensemble members near the phase vector $\mathbf{\Gamma}$ at time t.

There is a simple, exact equation of motion for this density. That equation is called the *phase continuity equation* (Gibbs, 1981). (In most textbooks, this equation is called the *Liouville equation*. In fact, Liouville's 1838 paper (Liouville, 1838) does not refer to the phase space density at all.)

The proof of the phase continuity equation

$$\frac{df(\mathbf{\Gamma}; t)}{dt} = -f(\mathbf{\Gamma}, t)\frac{\partial}{\partial \mathbf{\Gamma}} \cdot \dot{\mathbf{\Gamma}}(\mathbf{\Gamma}) = -f(\mathbf{\Gamma}, t)\Lambda(\mathbf{\Gamma}) \tag{2.26}$$

is identical to the proof of the mass continuity equation for a compressible fluid in hydrodynamics (Evans and Morriss, 1990). Both equations express the fact that the total mass of a compressible fluid or the total number of ensemble members in phase space is conserved. The total number $N_{\mathbf{\Gamma}}$ of ensemble members inside an enclosing phase space volume $V_{\mathbf{\Gamma}}$ must be related to the total integrated flux taken over the enclosing surface $S_{\mathbf{\Gamma}}$:

$$\begin{aligned}\frac{dN_{\Gamma}}{dt} &= \int_{V_{\Gamma}} d\mathbf{\Gamma}\frac{\partial f(\mathbf{\Gamma};t)}{\partial t} = -\int_{S_{\Gamma}} d\mathbf{S}_{\Gamma}\cdot\dot{\mathbf{\Gamma}}f(\mathbf{\Gamma};t)\\ &= -\int_{V_{\Gamma}} d\mathbf{\Gamma}\frac{\partial}{\partial\mathbf{\Gamma}}\cdot[\dot{\mathbf{\Gamma}}f(\mathbf{\Gamma};t)]\end{aligned} \tag{2.27}$$

Since this equation is true for arbitrary phase space volumes V_{Γ}, we see that

$$\begin{aligned}\frac{\partial f(\mathbf{\Gamma};t)}{\partial t} &= -\frac{\partial}{\partial\mathbf{\Gamma}}\cdot[\dot{\mathbf{\Gamma}}f(\mathbf{\Gamma};t)]\\ &= -f(\mathbf{\Gamma};t)\frac{\partial}{\partial\mathbf{\Gamma}}\cdot\dot{\mathbf{\Gamma}} - \dot{\mathbf{\Gamma}}\cdot\frac{\partial f(\mathbf{\Gamma};t)}{\partial\mathbf{\Gamma}}\\ &= -\left(\frac{\partial}{\partial\mathbf{\Gamma}}\cdot\dot{\mathbf{\Gamma}}+\dot{\mathbf{\Gamma}}\cdot\frac{\partial}{\partial\mathbf{\Gamma}}\right)f(\mathbf{\Gamma};t)\end{aligned} \tag{2.28}$$

where the chain rule is used to obtain line 2. For isokinetic or isoenergetic systems that have fixed total momentum in three Cartesian dimensions and satisfy AI$\mathbf{\Gamma}$, the phase space expansion factor Λ is (see Appendix 2.A):

$$\Lambda \equiv \frac{\partial}{\partial\mathbf{\Gamma}}\cdot\dot{\mathbf{\Gamma}} = -(3N_{\text{th}}-4)\alpha \tag{2.29}$$

This equation is exact for any arbitrary system size. It does not contain any large N approximations (see Appendix 2.A).

Definition

In order to carry out symbolic calculations of the time-dependent N-particle distribution function, it is convenient to define the *f-Liouvillean* L, where

$$\left(\frac{\partial}{\partial\mathbf{\Gamma}}\cdot\dot{\mathbf{\Gamma}}+\dot{\mathbf{\Gamma}}\cdot\frac{\partial}{\partial\mathbf{\Gamma}}\right)\cdots \equiv iL\cdots \tag{2.30}$$

(We note in passing that, while Liouville never discussed phase space density as required in the phase continuity equation, he did discuss, using very different notation (Liouville, 1838), what we call the p-Liouvillean.) Using this operator, we see that the phase continuity equation can be written as

$$\frac{\partial f(\mathbf{\Gamma};t)}{\partial t} = -iLf(\mathbf{\Gamma};t) \tag{2.31}$$

This equation has a formal solution

$$f(\mathbf{\Gamma};t) = \exp[-iLt]f(\mathbf{\Gamma};0) \tag{2.32}$$

The correctness of this solution can be checked by differentiation.

Proof of Eq. (2.26) is completed by using Eq. (2.28) to show

$$\begin{aligned}\frac{df(\boldsymbol{\Gamma};t)}{dt} &= \frac{\partial f(\boldsymbol{\Gamma};t)}{\partial t} + \dot{\boldsymbol{\Gamma}}(\boldsymbol{\Gamma})\cdot\frac{\partial f(\boldsymbol{\Gamma};t)}{\partial \boldsymbol{\Gamma}}\\ &= -\frac{\partial}{\partial \boldsymbol{\Gamma}}\cdot[\dot{\boldsymbol{\Gamma}}(\boldsymbol{\Gamma})f(\boldsymbol{\Gamma};t)] + \dot{\boldsymbol{\Gamma}}(\boldsymbol{\Gamma})\cdot\frac{\partial f(\boldsymbol{\Gamma};t)}{\partial \boldsymbol{\Gamma}}\\ &= -f(\boldsymbol{\Gamma};t)\frac{\partial}{\partial \boldsymbol{\Gamma}}\cdot\dot{\boldsymbol{\Gamma}}(\boldsymbol{\Gamma}) = -f(\boldsymbol{\Gamma};t)\Lambda(\boldsymbol{\Gamma}), \quad \forall\boldsymbol{\Gamma}\in D\end{aligned} \tag{2.33}$$

If we set $\boldsymbol{\Gamma} \to S^t\boldsymbol{\Gamma}$, we obtain

$$\frac{df(S^t\boldsymbol{\Gamma};t)}{dt} = -\Lambda(S^t\boldsymbol{\Gamma})f(S^t\boldsymbol{\Gamma};t) \tag{2.34}$$

Definition

Equation (2.34) is termed the *streaming* or *Lagrangian form of the phase continuity equation.* For a given initial phase $\boldsymbol{\Gamma}$, Eq. (2.34) is a simple first-order ordinary differential equation for the density along the phase space trajectory. Its solution can be written as

$$f(S^t\boldsymbol{\Gamma};t) = \exp\left[-\int_0^t ds\,\Lambda\,(S^s\boldsymbol{\Gamma})\right] f(\boldsymbol{\Gamma};0) \tag{2.35}$$

the correctness of which can easily be checked by differentiation: $df(S^t\boldsymbol{\Gamma};t)/dt = -\Lambda(S^t\boldsymbol{\Gamma})f(S^t\boldsymbol{\Gamma};t)$. So the distribution function at time t at the streamed position of the phase vector originating at $\boldsymbol{\Gamma}$ is related to the path integral of the phase space expansion factor along the phase space trajectory.

We could also consider the time dependence of the measure of an infinitesimal phase space volume $dV_{\Gamma}(S^s\boldsymbol{\Gamma})$ centered on the streamed position $S^s\boldsymbol{\Gamma} : 0 \le s \le t$ along the phase space trajectory. This phase space volume contains a fixed number of ensemble members and obeys the following equation of motion:

$$dV_{\Gamma}(S^t\boldsymbol{\Gamma}) = \exp\left[\int_0^t ds\,\Lambda\,(S^s\boldsymbol{\Gamma})\right] dV_{\Gamma}(\boldsymbol{\Gamma}) \tag{2.36}$$

In nonequilibrium steady states, experience shows that the time-averaged value of the thermostat multiplier is positive (we will prove this later using the fluctuation theorem) and therefore the time-averaged phase space expansion factor is negative. This implies that, for almost every initial phase space vector, the streamed density $f(S^t\boldsymbol{\Gamma};t) \sim \exp[(D_C(N_{\text{th}}-1))\overline{\alpha}t]f(\boldsymbol{\Gamma};0) \to +\infty,\ \ \forall!\boldsymbol{\Gamma}\in D$, while the corresponding streamed phase space volume goes to zero exponentially in time.

Definition

We are now in a position to compute the *ensemble averages* $\langle B(t)\rangle$ of an arbitrary integrable phase function $B(\boldsymbol{\Gamma})$:

$$\langle B(t)\rangle \equiv \int d\boldsymbol{\Gamma}\, B(\boldsymbol{\Gamma})f(\boldsymbol{\Gamma};t) = \int d\boldsymbol{\Gamma}\, B(S^t\boldsymbol{\Gamma})f(\boldsymbol{\Gamma};0) \tag{2.37}$$

Physically, this equation is rather obvious. We can formally prove the correctness of this equation by noting that by integrating by parts (see Evans and Morriss (1990, Eq. (3.53)):

$$\int d\mathbf{\Gamma} f(\mathbf{\Gamma};t) iLB(\mathbf{\Gamma}) = -\int d\mathbf{\Gamma}\, B(\mathbf{\Gamma}) i\mathcal{L} f(\mathbf{\Gamma};t) \tag{2.38}$$

From the above equation, we deduce that

$$\begin{aligned}\langle B(t)\rangle &= \int d\mathbf{\Gamma}\, B(\mathbf{\Gamma}) f(\mathbf{\Gamma};t)\\ &= \int d\mathbf{\Gamma}\, B(\mathbf{\Gamma}) \exp[-i\mathcal{L}t] f(\mathbf{\Gamma};0)\\ &= \int d\mathbf{\Gamma} f(\mathbf{\Gamma};0) \exp[iLt] B(\mathbf{\Gamma})\\ &\equiv \int d\mathbf{\Gamma} f(\mathbf{\Gamma};0) B(S^t\mathbf{\Gamma})\end{aligned} \tag{2.39}$$

We note that, if the system satisfies the AI$\mathbf{\Gamma}$ condition and if there are no thermostats applied, then $\Lambda = 0$ and $L = \mathcal{L}$; that is, the Liouville operator is Hermitian. For thermostatted systems, the *f*- and *p*-Liouvilleans are not equal.

Finally, we note that all phase space integrals given above should be carried out over some specified phase space domain. We have omitted this for simplicity, but if one wants to verify these equations, for example, Eq. (2.38) by integrating by parts, then the phase space domain is important.

2.5 Lyapunov Instability and Statistical Mechanics

In this section we give the briefest of introductions (Evans and Searles, 2002) to a relatively new field of research. Many statements are made without proof. The interested reader should consult the references cited in this section.

We include this material because without some knowledge of the dimensional reduction processes in time-reversible deterministic steady states, the reader will be puzzled by many apparent contradictions. How can it be that in a nonequilibrium *steady state* the Gibbs entropy is not time-independent but instead decreases at a constant average *rate* toward negative infinity?

The Lyapunov exponents are used in dynamical systems theory to characterize the stability of phase space trajectories. If one imagines two systems that evolve in time from phase vectors $\mathbf{\Gamma}_1, \mathbf{\Gamma}_2$ that are infinitesimally separated at time zero, $|\mathbf{\Gamma}_1 - \mathbf{\Gamma}_2| \equiv |\delta\mathbf{\Gamma}_1| \to 0$, then one can ask how the separation between these two systems evolves in time. Oseledec's theorem (see Evans and Morriss, 1990; Ruelle, 1979) says that for nonintegrable systems under very general conditions the separation vector asymptotically grows or shrinks *exponentially* in time. This does not happen for integrable systems, but very few physical systems are integrable. A system is said to be chaotic if the separation vector asymptotically *grows* exponentially with time. Most systems in Nature are chaotic: the world weather and high Reynolds number flows are chaotic. In fact, all systems that obey thermodynamics are chaotic.

In 1990, the first of a remarkable set of relationships between phase space stability measures (i.e., Lyapunov exponents) and thermophysical properties was discovered by Evans, Cohen, and Morriss (1990) and, separately and differently, by Gaspard and Nicolis (1990). More recently, Lyapunov exponents have been used to assign dynamical probabilities to the observation of phase space trajectory segments (Gallavotti and Cohen, 1995a,b). This is something quite new to statistical mechanics, where hitherto probabilities had been given (only for equilibrium systems!) on the basis of the value of the Hamiltonian (i.e., the weights are static).

Suppose the autonomous equations of motion are written as

$$\dot{\boldsymbol{\Gamma}} = \dot{\boldsymbol{\Gamma}}(\boldsymbol{\Gamma}) \tag{2.40}$$

Definition

It is trivial to see that the equation of motion for an infinitesimal phase space separation vector $d\boldsymbol{\Gamma}$ can be written as:

$$d\dot{\boldsymbol{\Gamma}} = \mathbf{T}(\boldsymbol{\Gamma})d\boldsymbol{\Gamma} \tag{2.41}$$

where $\mathbf{T} \equiv \partial\dot{\boldsymbol{\Gamma}}(\boldsymbol{\Gamma}, t)/\partial\boldsymbol{\Gamma}$ is the *stability matrix* for the flow.

Definition

The propagation of the tangent vectors is therefore given by

$$d(S^t\boldsymbol{\Gamma}) = \boldsymbol{\Xi}(\boldsymbol{\Gamma}; t) \cdot d\boldsymbol{\Gamma} \tag{2.42}$$

where the *tangent vector propagator* is

$$\boldsymbol{\Xi}(\boldsymbol{\Gamma}; t) = \exp_L\left(\int_0^t ds\,\mathbf{T}\,(S^s\boldsymbol{\Gamma})\right) \tag{2.43}$$

and $\exp_L$ is a left time-ordered exponential (i.e., the sum in the exponent is time-ordered with the latest times $\exp\{\lim_{\Delta t\to 0}\Delta t[\mathbf{T}(S^t\boldsymbol{\Gamma}) + \mathbf{T}(S^{t-\Delta t}\boldsymbol{\Gamma}) + \cdots + \mathbf{T}(\boldsymbol{\Gamma})]\}$ at the left). We note that the stability matrices evaluated at different times do not commute. The correctness of Eq. (2.42) can be checked by differentiation.

The number of Lyapunov exponents is equal to the dimension of the ostensible phase space of the system. There are two ways of specifying this when there are constants of the motion. In one case, the ostensible phase space can be considered to be the full $2D_CN$-dimensional space, where D_C is the Cartesian dimension of the phase space and N is the number of particles, and although motion occurs in a subspace of this, perturbations off that subspace (which do not satisfy the constraints) are allowed in consideration of the Lyapunov spectrum or the phase space expansion.

Another way is to assume that the ostensible phase is the hyperspace within that $2D_CN$ space defined by those constants of motion. Then perturbations off this hypersurface are not considered in the determination of the Lyapunov spectrum or the phase space expansion. Of course, since the equations of motion of the system remain the same irrespective of this treatment, this choice is irrelevant for the physical properties of the system. However, it is relevant for properties such as the Lyapunov exponents and phase space expansion. Selection of the approach used is often based on the convenience of the mathematical or numerical analysis. In this section on the Lyapunov exponents, it is more convenient to consider the full $2D_CN$-dimensional space, although in other parts of the book the reduced dimension is used.

If $d\mathbf{\Gamma}_i$ is an eigenvector of $\Xi^T(\mathbf{\Gamma};t)\Xi(\mathbf{\Gamma};t)$ and if the Lyapunov exponents are defined as

$$\left\{\lambda_i;\, i=1,\ldots,2D_CN\right\} \equiv \lim_{t\to\infty}\frac{1}{2t}\ln\left(\text{eigenvalues}\left(\Xi^T(\mathbf{\Gamma};t)\,\Xi(\mathbf{\Gamma};t)\right)\right) \tag{2.44}$$

then the Lyapunov exponents describe the growth rates of the set of orthogonal tangent vectors (eigenvectors of $(\Xi^T(t)\Xi(t))$), $\{d\mathbf{\Gamma}_i(t); i=1,2D_CN\}$,

$$\begin{aligned}\lim_{t\to\infty}\frac{1}{2t}\ln\frac{\left|d\left(S^t\mathbf{\Gamma}_i\right)\cdot d(S^t\mathbf{\Gamma}_i)\right|}{\left|d\mathbf{\Gamma}_i\cdot d\mathbf{\Gamma}_i\right|} &= \lim_{t\to\infty}\frac{1}{2t}\ln\frac{\left|d\mathbf{\Gamma}_i^{\,T}\Xi^T(t)\,\Xi(t)d\mathbf{\Gamma}_i\right|}{\left|d\mathbf{\Gamma}_i\cdot d\mathbf{\Gamma}_i\right|}\\ &= \frac{1}{2t}\ln\frac{\left|d\mathbf{\Gamma}_i^{\,T}\exp\left[2\lambda_i t\right]d\mathbf{\Gamma}_i\right|}{\left|d\mathbf{\Gamma}_i\cdot d\mathbf{\Gamma}_i\right|}\\ &= \lambda_i,\quad i=1,\ldots,2D_CN\end{aligned} \tag{2.45}$$

(Note: the transpose matrix not only transposes the rows and column but also transposes time ordering $\Xi^T(\mathbf{\Gamma};t)=\exp_R\left(\int_0^t ds\,\mathbf{T}^T\left(S^s\mathbf{\Gamma}\right)\right)$.)

By convention, the exponents are ordered such that $\lambda_1\geq\lambda_2\geq\cdots\geq\lambda_{2D_CN}$. It can be shown that the Lyapunov exponents are independent of the metric used to measure phase space lengths. They are also independent, for T-mixing steady states (see Chapter 6), of the initial position $\mathbf{\Gamma}$ of the "mother" phase space trajectory.

In general, there will be a number of Lyapunov exponents that are zero. For example, in an equilibrium system, there will $2D_C$ zero exponents for each Cartesian momentum component that is conserved because momentum conservation also means that the associated position of the center of mass of each Cartesian coordinate is constant. In autonomous systems, there will be another zero exponent associated with time translation invariance. In isokinetic or isoenergetic systems, there will each be another zero exponent associated with this additional constant of the motion. To keep the notation flexible, we will say that there are f zero Lyapunov exponents.

In order to calculate the Lyapunov spectrum, one does not normally use Eq. (2.44). Benettin *et al.* developed a technique (Benettin, Galgani, and Strelcyn, 1976) wherein the finite but small displacement vectors are periodically rescaled and orthogonalized during the course of a solution of the equations of motion. Hoover and Posch (1985) and, independently, Goldhirsch, Sulem, and Orszag (1987) pointed out that this rescaling and orthogonalization can be carried out continuously by introducing constraints to the equations of motion of the tangent vectors. With this modification, orthogonality and tangent vector lengths are maintained at all times during the calculation in much the same way as our Gaussian thermostats and ergostats maintain fixed values for the kinetic temperature or the internal energy.

In theory, the $2D_CN$ eigenvalues of the real, symmetric matrix $\Xi^T(t) \cdot \Xi(t)$ can also be used to calculate the Lyapunov spectrum in the limit $t \to \infty$. Since $\Xi(S^t\boldsymbol{\Gamma})$ is dependent only on the mother trajectory $S^t\boldsymbol{\Gamma}$, calculation of the Lyapunov exponents from the eigenvalues of $\Xi^T(t) \cdot \Xi(t)$ does not require the solution of $2D_CN$ tangent trajectories as in the methods mentioned in the previous paragraph. However, after a short time, numerical difficulties are encountered using this method because of the enormous difference in the magnitude of the eigenvalues of the $\Xi^T(t) \cdot \Xi(t)$ matrix. After a short time, this matrix becomes highly ill-conditioned. The use of QR decompositions (where $\Xi(t) = \mathbf{Q} \cdot \mathbf{R}$ and $\mathbf{R}$ is a real upper triangular matrix with positive diagonal elements and $\mathbf{Q}$ is a real orthogonal matrix) reduces this problem. Use of the QR decomposition is equivalent to the reorthogonalization/rescaling of the displacement vectors in the scheme discussed previously.

We note that the Lyapunov exponents are only defined in the long time limit and, if the simulated *nonequilibrium* fluid does not reach a stationary state, the exponents will not converge to constant values. It is useful for the purposes of this work to define time-dependent exponents as

$$\left\{\lambda_i(t;\boldsymbol{\Gamma});\, i=1,\ldots,2D_CN\right\} = \frac{1}{2t}\ln\left(\text{eigenvalues}\left(\Xi^T(t;\boldsymbol{\Gamma})\cdot\Xi(t;\boldsymbol{\Gamma})\right)\right) \quad (2.46)$$

Unlike the Lyapunov exponents, these finite time exponents will depend on the initial phase space vector $\boldsymbol{\Gamma}$ and the length of time over which the tangent vectors are integrated, and we therefore will refer to them as finite-time, local Lyapunov exponents.

The systems considered in statistical thermodynamics are all chaotic: they have at least one positive Lyapunov exponent. This means that (except for a set of zero measure) points that are initially close will diverge after some time, and therefore information on the initial phase space position of the trajectory will be lost. Points that are initially close will eventually span the accessible phase space of the system. In fact, for typical macroscopic nonequilibrium systems, the number of positive and negative exponents are very nearly equal. The total number of exponents is, of course, of the order of Avodagro's constant ($\sim 6 \times 10^{23}$).

The sum of the first two Lyapunov exponents shows how quickly the fastest growing area grows. The sum of the first three Lyapunov exponents gives the rate of growth of the fastest growing three-dimensional volume, and so on. The Lyapunov exponents of an equilibrium (Hamiltonian) system sum to zero, reflecting the phase space conservation of these systems.

The sum of all the Lyapunov exponents is, in fact, equal to the time-averaged phase space expansion factor, so that for isokinetic systems

$$\sum_{i=1}^{2D_C N} \lambda_i = \lim_{t\to\infty} \overline{\frac{\partial}{\partial \mathbf{\Gamma}} \cdot \dot{\mathbf{\Gamma}}}_t = \lim_{t\to\infty} \overline{\Lambda}_t = -\lim_{t\to\infty} [D_C(N_{\text{th}} - 1) - 1]\overline{\alpha}_t \tag{2.47}$$

This is because the phase space expansion is the average rate of increase of the ostensibly dimensioned phase space volume element. Comparing this equation with Eq. (2.36) shows that the sum of all the Lyapunov exponents gives the exponential rate at which the streamed phase space volume vanishes.

$$\lim_{t\to\infty} \delta V_{\mathbf{\Gamma}}(S^t\mathbf{\Gamma}) = \exp[\overline{\Lambda}_t t]\delta V_{\mathbf{\Gamma}}(\mathbf{\Gamma}) = \exp\left[\sum_{i=1} \lambda_i t\right] \delta V_{\mathbf{\Gamma}}(\mathbf{\Gamma}) \tag{2.48}$$

As we will see later, in equilibrium systems all properties, including Lyapunov exponents, must be invariant under time reversal. This implies that time reversal of Lyapunov spectra for equilibrium systems must transform the spectrum into itself, which in turn means that for all equilibrium systems the exponents must arrange themselves into conjugate pairs that each sum to zero.

$$\lambda_{\max}^{\text{eq}} + \lambda_{\min}^{\text{eq}} = \lambda_{\max+1}^{\text{eq}} + \lambda_{\min-1}^{\text{eq}} = \cdots = 0 \tag{2.49}$$

If the ostensible phase space dimension is odd, the unpaired exponent must be zero. In fact, there could be multiple exponents that are zero since this would not violate the time-reversal property.

The symplectic eigenvalue theorem (Eckmann and Ruelle, 1985; Abraham and Marsden, 1978) shows that, for all autonomous symplectic dynamical systems with time-independent Lyapunov exponents, the conjugate exponents must pair about zero as in Eq. (2.49). If the system is stationary in time, this pairing about zero can happen only if the system eventually relaxes toward equilibrium. As we will see later (Section 4.3), the necessary and sufficient condition for stationarity is that the system is ΩT-mixing. Hence we have our first proof of the relaxation to equilibrium. All autonomous symplectic systems that are ΩT-mixing *must* relax to equilibrium at long times. Symplectic systems include autonomous Hamiltonian systems as a special case.

Equation (2.49) also shows how Lyapunov exponents are related to time-averaged dissipative fluxes and thereby to transport coefficients. We define a nonlinear transport coefficient in terms of the long time average of the dissipative flux divided by the dissipative field as

$$L(F_e) \equiv -\lim_{t\to\infty} \frac{\overline{J}_t}{F_e} \tag{2.50}$$

Definition

Then using equations we derive what is known as the *Lyapunov sum rule*, which for an N-particle isokinetic system reads

$$L_N(F_e) = -\frac{k_B T \sum_{i=1}^{2D_C N} \lambda_i}{F_e^2 V} \tag{2.51}$$

For homogeneously thermostatted systems that are adiabatically symplectic, Evans, Cohen, and Morriss (1990) have shown that the Lyapunov spectrum has a conjugate pairing symmetry about the time-averaged value of the thermostat multiplier so that the nonlinear transport coefficient can be calculated by summing *any* conjugate pair of Lyapunov exponents. Since the largest and the smallest exponents are the easiest to compute, for such systems we can write

$$L_N(F_e) = -\frac{k_B T\left(\lambda_{\max} + \lambda_{\min}\right)\left(2D_C N - f\right)/2}{F_e^2 V} \tag{2.52}$$

Definition

Equation (2.52) is an example of the *conjugate pairing rule* for homogeneously thermostatted, adiabatically symplectic systems.

Equations (2.51) and (2.52) show how apparently abstract mathematical quantities such as Lyapunov exponents, which characterize the stability or otherwise of phase space trajectories, are related to measurable physical properties such as transport coefficients. In physical systems of thermodynamic interest, the Lyapunov spectra are very smooth.

For thermostatted steady states, the Lyapunov sum is negative. This indicates that the phase space collapses toward a lower dimensional attractor in the original phase space. A number of relationships have been conjectured to characterize the dimension of this object whose volume is preserved by the dynamics (the so-called invariant steady-state attractor).

Definition

For systems of interacting particles, this dimension is commonly taken to be the *Kaplan–Yorke dimension* defined as (Kaplan and Yorke, 1979; Frederickson, Kaplan, and Yorke, 1983)

$$D_{KY,N} \equiv N_{KY} + \sum_{i=1}^{N_{KY}} \lambda_i / \left|\lambda_{N_{KY}+1}\right| \tag{2.53}$$

where $N_{N_{KY}}$ is the largest integer for which $\sum_{i=1}^{N_{KY}} \lambda_i > 0$. As you sum the Lyapunov exponents from the largest to the smallest, you start by summing at least one positive number – because the system is chaotic. If the system satisfies the second "law" of thermodynamics, the time-averaged thermostat multiplier is positive, indicating that, on average, work is converted to heat that must be removed by the thermostat in order to maintain steady state conditions. From the Lyapunov sum rule, that is, Eq. (2.52), we see that summing all the Lyapunov exponents gives a negative number. Somewhere during the summation process, the running sum changed from being positive to negative. Using linear interpolation between the integer exponents, the Kaplan–Yorke dimension is determined and is considered to be a measure of the dimension of that object whose volume is preserved by the dynamics. In typical systems of thermophysical interest, the ostensible dimension of phase space is $O(6N_A)$ where N_A is Avogadro's number. The typical dimensional reduction is, by way of contrast, exceedingly small $O(1)$! In the linear response regime close to equilibrium, there is an exact relation between the Kaplan–Yorke dimension of the steady state and the zero-field transport coefficient for an N-particle system (Evans *et al.*, 2000):

$$L_N(F_e = 0) = \lim_{F_e \to 0} \frac{(2D_C N - f - D_{KY,N}(F_e))\lambda_{\max}(F_e) k_B T}{V F_e^2} \tag{2.54}$$

In this limit, the dimensional reduction is less than 1: $2D_C N - f - D_{KY,N}(F_e) < 1$.

These remarkable equations show not only how to calculate the Kaplan–Yorke dimension of the invariant steady state attractor but also how this dimensional reduction is related to a physical property, namely a transport coefficient.

It is important to remember that, although we can use the Kaplan–Yorke construction to determine the dimension of an invariant mathematical object, the phase continuity equation for nonequilibrium steady states shows that on average the $6N$-dimensional volume occupied by an ensemble of phase space trajectories contracts *forever* (see Eq. (2.36)) and is never stationary no matter how long the trajectories are run. The Kaplan–Yorke construction is just a mathematical construction. Important physical information is conveyed in the rate of collapse toward the zero volume attractor. In nonequilibrium steady states, this collapse continues forever. The reduced dimension of the nonequilibrium steady state only has consequences for some phase functions.

Definition

We define a physical phase function $A(\boldsymbol{\Gamma})$ to be any physical property that can be written as a function of the phase space vector $\boldsymbol{\Gamma}$. Examples include the internal energy $H_0(\boldsymbol{\Gamma})$ and the pressure tensor $\mathbf{P}(\boldsymbol{\Gamma}) = \left[\sum_{i\in V}\mathbf{p}_i\mathbf{p}_i + {}^1\!/_2\sum_{i,j}\mathbf{F}_{ij}\left(\mathbf{q}_i - \mathbf{q}_j\right)\right]/V$ (for bulk systems with pair interactions only). These physical phase functions have meaning for an individual microstate. For autonomous Hamiltonian systems, the internal energy in a co-moving (or nonmoving) reference frame is a constant of the motion: $dH_0(\boldsymbol{\Gamma})/dt = \dot{\boldsymbol{\Gamma}} \cdot \partial H_0(\boldsymbol{\Gamma})/\partial\boldsymbol{\Gamma} = 0$. For adiabatic SLLOD dynamics, the xy element of the pressure tensor is the dissipative flux, and satisfies Eq. (2.20) *instantaneously* for an individual mechanical N-particle system: $\dot{H}_0(\boldsymbol{\Gamma}) = -P_{xy}(\boldsymbol{\Gamma})\dot{\gamma}(t)V$!

Of course, one can calculate the average value of these physical phase functions for a macroscopic ensemble of systems: $\langle A(t)\rangle \equiv \int d\boldsymbol{\Gamma} A(\boldsymbol{\Gamma}) f(\boldsymbol{\Gamma};t)$. When this is done, the averages of physical phase functions can be expressed as functionals of low order distribution functions (e.g., the singlet, pair, or three-body distribution functions (Evans and Rondoni, 2002), $f^{(1)}(\mathbf{q},\mathbf{p})$, $f^{(2)}(\mathbf{q}_1, \mathbf{q}_2, \mathbf{p}_1, \mathbf{p}_2)$, $f^{(3)}(\mathbf{q}_1, \mathbf{q}_2, \mathbf{q}_3, \mathbf{p}_1, \mathbf{p}_2, \mathbf{p}_3)$, etc.). Without ensemble averaging, the value of these physical phase functions gives the value of the physical property for an individual microstate (e.g., its *instantaneous* energy, stress kinetic temperature or pressure). This is what we mean when we use the description "physical phase function."

In contrast, the Gibbs entropy $-k_B \int d\boldsymbol{\Gamma} f(\boldsymbol{\Gamma};t)\ln(f^{(N)}(\boldsymbol{\Gamma};t)$ is a functional of the full N-particle distribution function $f(\boldsymbol{\Gamma};t) \equiv f^{(N)}(\mathbf{q},\mathbf{p};t)$, which cannot be expressed in terms of only the low order distribution functions. A criterion for whether a phase function is high or low order can be made on the basis of whether it is a functional of distributions that span a space whose dimension is higher or lower than the Kaplan–Yorke dimension.

As mentioned previously, in real experimental nonequilibrium steady states the Kaplan–Yorke dimension of the steady state attractor is typically very close to the ostensible dimension of phase space. In real situations, the dimensional reductions are tiny (Evans *et al.*, 2000). Low order phase functions are formed by integrating over many of the coordinates and momenta of the N-particle distribution function. This smooths any singularities present in the high order distributions. The low order distributions do not notice that the nonequilibrium steady state phase space distribution is fractal. Only high order functions like the entropy and the various free energies are affected by the fractal singularities.

Unlike the case for physical phase variables, you cannot calculate the contribution an individual microstate makes to the entropy $f(\mathbf{\Gamma};t)\ln(f(\mathbf{\Gamma};t))$ without knowledge of the full N-particle phase space distribution, $f(\mathbf{\Gamma};t)$. The phase space probability density is a property that can be determined *only* from knowledge of the *entire* macroscopic ensemble. Unlike physical phase functions, there is no microscopic meaning that can be attached to the Gibbs entropy of an individual microstate at phase space vector $\mathbf{\Gamma}$ at time t. In this sense, Gibbs entropy is not a physical property of a mechanical microstate. It is, in fact, a property of the ensemble of microstates.

2.6
Gibbs Entropy in Deterministic Nonequilibrium Macrostates

Definition

The *fine-grained Gibbs entropy* S_G is defined as

$$S_G(t) \equiv -k_B \int_D d\mathbf{\Gamma} f(\mathbf{\Gamma};t) \ln\left[f(\mathbf{\Gamma};t)\right] \tag{2.55}$$

See Section 5.6 and Chapter 8 for discussions on why this quantity is useful at equilibrium. At this stage we do not discuss why entropy is defined in this way. As mentioned above, the Gibbs entropy is obviously a functional of the full N-particle distribution function.

We will, however, explain why this quantity is so problematic in deterministic nonequilibrium steady states (Evans, Williams, and Searles, 2011). We know that the nonequilibrium density is distributed in a space of lower dimension than the ostensible phase space dimension: for example, $2D_C(N_{th}-1)-1$ for a system with a Gaussian thermostat or ergostat. This is analogous to condensing a density from a two-dimensional area onto a one-dimensional line. When the entropy is defined as above, the integral is to be taken over the ostensible phase space D. This is highly problematic because almost everywhere in the phase space the density measured with respect to the ostensible dimension is zero!

The fact is that you can only calculate the entropy if you know the dimension and the topology of the measure. However, the dimension is known only approximately and the topology is, in general, not known at all.

If we use the phase continuity equation, we can attempt to calculate how this fine-grained entropy changes in time for an isokinetic system:

$$\begin{aligned}
\dot{S}_G(t) &= -k_B \int d\mathbf{\Gamma}\,[1+\ln(f)]\frac{\partial f}{\partial t} \\
&= k_B \int d\mathbf{\Gamma}\,[1+\ln(f)]\frac{\partial}{\partial \mathbf{\Gamma}}\cdot[\dot{\mathbf{\Gamma}}f] \\
&= -k_B \int d\mathbf{\Gamma}\, f\dot{\mathbf{\Gamma}}\cdot\frac{\partial}{\partial \mathbf{\Gamma}}[1+\ln(f)] \\
&= -k_B \int d\mathbf{\Gamma}\,\dot{\mathbf{\Gamma}}\cdot\frac{\partial f}{\partial \mathbf{\Gamma}} \\
&= k_B \int d\mathbf{\Gamma}\, f(\mathbf{\Gamma};t)\frac{\partial}{\partial \mathbf{\Gamma}}\cdot\dot{\mathbf{\Gamma}}(\mathbf{\Gamma},t) = -k_B(D_C(N_{\text{th}}-1)-1)\langle\alpha(t)\rangle
\end{aligned} \tag{2.56}$$

Integration by parts is used in going from the second to the third line, with the boundary terms equaling zero. In a nonequilibrium steady state, $\langle\alpha(t)\rangle = \text{const} > 0$ (the sign of the average can be proved from the fluctuation theorem, see Section 3.2 and noting the relationship between the average of the dissipation function and the thermostat multiplier), so the entropy apparently diverges at a constant rate toward negative infinity!

If there are no thermostats, as in an autonomous Hamiltonian system, the fine-grained Gibbs entropy is simply constant because $\dot{\mathbf{\Gamma}}\cdot\partial/\partial\mathbf{\Gamma} = 0$. So for autonomous Hamiltonian systems, the entropy, which only exists at the level of a macroscopic ensemble, is a constant of the motion as the entire ensemble evolves in time. That ensemble consists of an infinite set of noninteracting N-particle systems. Entropy is not a physical property that can be ascribed to an individual N-particle microstate.

One may object that in Eq. (2.56) the integrations by parts involve boundary terms that may be nonzero. This is not the case, however. If we consider an autonomous Hamiltonian system that has some arbitrary initial distribution, we can show that the Gibbs entropy is time-independent directly rather than showing that its time derivative is zero.

We know from the streaming form of the phase continuity equation (2.26) that $f(S^t\mathbf{\Gamma};t) = f(\mathbf{\Gamma};0)$. This is because $f(S^s\mathbf{\Gamma};t) = \exp\left[-\int_0^t ds\,\Lambda(S^s\mathbf{\Gamma})\right]f(\mathbf{\Gamma};0)$ and for Hamiltonian dynamics $\Lambda(\mathbf{\Gamma}) \equiv 0$. We use Eq. (2.55) to calculate the entropy at some time t.

$$\begin{aligned}
S_G(t) &= -k_B \int d(S^t\mathbf{\Gamma}) f(S^t\mathbf{\Gamma};t)\ln[f(S^t\mathbf{\Gamma};t)] \\
&= -k_B \int d(S^t\mathbf{\Gamma}) f(\mathbf{\Gamma};0)\ln[f(\mathbf{\Gamma};0)] \\
&= -k_B \int d\mathbf{\Gamma}\left|\frac{\partial\left(S^t\mathbf{\Gamma}\right)}{\partial\mathbf{\Gamma}}\right| f(\mathbf{\Gamma};0)\ln[f(\mathbf{\Gamma};0)] \\
&= -k_B \int d\mathbf{\Gamma}\, f(\mathbf{\Gamma};0)\ln[f(\mathbf{\Gamma};0)] = S_G(0)
\end{aligned} \tag{2.57}$$

The first line uses $S^t\mathbf{\Gamma}$ as a dummy integration variable in Eq. (2.55). The second line uses the streaming phase continuity relation for Hamiltonian systems. The fourth line uses the fact that the Jacobian in line 3 is, for Hamiltonian systems, unity.

The constancy of the entropy for autonomous Hamiltonian systems is almost never commented on in textbooks for physics, chemistry, or chemical engineering students. It is discussed in more mathematical texts on dynamical systems theory. This constancy was first discovered by Gibbs in 1903 (Gibbs, 1981). It was commented on extensively in the exhaustive, but ultimately inconclusive, Ehrenfest review of the foundations of statistical mechanics (Ehrenfest and Ehrenfest, 1990).

Because the entropy is so problematic in nonequilibrium systems, it will play no role in our discussions of nonequilibrium phenomena. We will meet it again when we discuss equilibrium systems in Chapter 8.

2.A Appendix: Phase Space Expansion Calculation

Here we consider the slightly tricky issue of computing the exact phase space expansion factor for Gaussian isokinetic dynamics. We treat the isokinetic case because it is less straightforward than the Nosé–Hoover case.

As discussed above, there are two ways of considering phase space expansion when there are constants of the motion: using the full $2D_C N$ dimensional space, where D_C is the Cartesian dimension of phase space and N is the number of particles, or to assume that the ostensible phase is the hyperspace within that $2D_C N$ space defined by those constants of motion. In the treatment below, we consider the latter choice. As an example of a case where the former case is considered for calculation of the phase space expansion, see Bright, Evans, and Searles (2005).

For simplicity, consider a three-dimensional N-particle system obeying the following dynamics:

$$\begin{aligned} \dot{\mathbf{q}}_i &= \frac{\mathbf{p}_i}{m} \\ \dot{\mathbf{p}}_i &= \mathbf{F}_i(\mathbf{q}) - \alpha \mathbf{p}_i \\ \alpha &= \frac{\sum_{i=1}^{N} \mathbf{F}_i \cdot \mathbf{p}_i}{\sum_{i=1}^{N} p_i^2} \end{aligned} \tag{2.A1}$$

As always, the momenta are peculiar so

$$\sum_{i=1}^{N} \mathbf{p}_i = \mathbf{0} \tag{2.A2}$$

With this choice for the thermostat multiplier, the peculiar kinetic energy is also constant:

$$\sum_{i=1}^{N} p_i(t)^2/2m = K, \quad \forall t \tag{2.A3}$$

The four constraints Eqs. (2.A2) and (2.A3) imply that the $3N$ Cartesian momentum components are not all independent, so one cannot compute the usual phase space expansion factor:

$$\Lambda = \partial/\partial\mathbf{\Gamma} \cdot \dot{\mathbf{\Gamma}}(\mathbf{\Gamma}) = \sum_{i=1}^{N} \partial/\partial\mathbf{p}_i \cdot (-\alpha\mathbf{p}_i) \tag{2.A4}$$

where we have assumed AI$\mathbf{\Gamma}$. The difficulty is that, in general, you cannot vary one Cartesian momentum component keeping all other $3N - 1$ components fixed and still satisfy Eq. (2.A2).

We resolve this situation by effectively eliminating the degrees of freedom associated with the Nth particle and compute the phase space expansion factor as

$$\begin{aligned}
\Lambda &= \sum_{i=1}^{N-1} \partial/\partial\mathbf{p}_i \cdot (-\alpha\mathbf{p}_i) \\
&= -(3N-3)\alpha - \sum_{i=1}^{N-1} \mathbf{p}_i \cdot \partial/\partial\mathbf{p}_i \alpha \\
&= -(3N-3)\alpha - \sum_{i=1}^{N-1} \mathbf{p}_i \cdot \partial/\partial\mathbf{p}_i \frac{\sum_{j=1}^{N-1} \mathbf{F}_j \cdot \mathbf{p}_j + \left(\sum_{j=1}^{N-1} \mathbf{F}_j\right) \cdot \left(\sum_{j=1}^{N-1} \mathbf{p}_j\right)}{\sum_{j=1}^{N-1} p_j^2 + \left(\sum_{j=1}^{N-1} \mathbf{p}_j\right)^2} \\
&= -(3N-3)\alpha - \frac{\sum_{i=1}^{N-1} \mathbf{F}_i \cdot \mathbf{p}_i + \left(\sum_{i=1}^{N-1} \mathbf{F}_j\right) \cdot \left(\sum_{i=1}^{N-1} \mathbf{p}_i\right)}{2mK} \\
&\quad + 2\frac{\sum_{i=1}^{N-1} \mathbf{F}_i \cdot \mathbf{p}_i + \left(\sum_{j=1}^{N-1} \mathbf{F}_j\right) \cdot \left(\sum_{j=1}^{N-1} \mathbf{p}_i\right)}{(2mK)^2} \left[\sum_{i=1}^{N-1} p_i^2 + \left(\sum_{j=1}^{N-1} \mathbf{p}_i\right)^2\right] \\
&= -(3N-4)\alpha
\end{aligned} \tag{2.A5}$$

In calculating this derivative, we still have one constraint. However the two terms involving the partial derivatives, namely $\sum_{i=1}^{N-1} \partial/\partial\mathbf{p}_i \cdot \mathbf{p}_i$ and $\sum_{i=1}^{N-1} \mathbf{p}_i \cdot \partial/\partial\mathbf{p}_i$, are independent of the value of the peculiar kinetic energy. So although the virtual displacement taken in the derivative violates the kinetic energy constraint, the answer that is computed is independent of the value of the kinetic energy. In fact, one could transform to a normalized momentum $\mathbf{p}_i'$ for which the scaled kinetic energy could not vary. The results so obtained are still given by Eq. (2.A5) because $\sum_{i=1}^{N-1} \partial/\partial \mathbf{p}_i \cdot \mathbf{p}_i = \sum_{i=1}^{N-1} \partial/\partial \mathbf{p}'_i \cdot \mathbf{p}'_i$, and so on.

The same calculation for Nosé–Hoover thermostats in the phase space extended to include the thermostat multiplier α shows that in that case the phase space expansion factor is $-(3N-3)\alpha$ because the second term in the second line of Eq. (2.A5) is absent.

References

Abraham, R. and Marsden, J.E. (1978) *Foundations of Mechanics*, Benjamin/Cummings, London.

Benettin, G., Galgani, L., and Strelcyn, J.-M. (1976) Kolmogorov entropy and numerical experiments. *Phys. Rev. A*, **14**, 2338–2345.

Bernardi, S., Brookes, S.J., and Searles, D.J. (2014) System size effects on calculation of the viscosity of extended molecules. *Chem. Eng. Sci.*, **121**, 236–244.

Boltzmann, L. (1872) Further studies on thermal equilibrium among Gas molecules. *Akad. Wiss. Wien*, **66**, 275.

Bright, J.N., Evans, D.J., and Searles, D.J. (2005) New observations regarding deterministic, time-reversible thermostats and Gauss's principle of least constraint. *J. Chem. Phys.*, **122**, 194106.

Daivis, P.J. and Todd, B.D. (2006) A simple, direct derivation and proof of the validity of the SLLOD equations of motion for generalized homogeneous flows. *J. Chem. Phys.*, **124**, 194103.

Eckmann, J.-P. and Ruelle, D. (1985) Ergodic theory of chaos and strange attractors. *Rev. Mod. Phys.*, **57**, 617.

Ehrenfest, P. and Ehrenfest, T. (1990) *The Conceptual Foundations of the Statistical Approach to Statistical Mechanics*, Dover, Mineola, NY.

Evans, D.J. (1983) Computer experiment for Non-linear thermodynamics of couette-flow. *J. Chem. Phys.*, **78**, 3297–3302.

Evans, D.J. (1985) Response theory as a free-energy extremum. *Phys. Rev. A*, **32**, 2923–2925.

Evans, D.J., Cohen, E.G.D., and Morriss, G.P. (1990) Viscosity of a simple fluid from its maximal Lyapunov exponents. *Phys. Rev. A*, **42**, 5990.

Evans, D.J., Cohen, E.G.D., Searles, D.J., and Bonetto, F. (2000) Note on the Kaplan-Yorke dimension and linear transport coefficients. *J. Stat. Phys.*, **101**, 17–34.

Evans, D.J., Hoover, W.G., Failor, B.H., Moran, B., and Ladd, A.J.C. (1983) Nonequilibrium molecular-dynamics Via gauss principle of least constraint. *Phys. Rev. A*, **28**, 1016–1021.

Evans, D.J. and Morriss, G.P. (1984a) Non-Newtonian molecular-dynamics. *Comput. Phys. Rep.*, **1**, 297–343.

Evans, D.J. and Morriss, G.P. (1984b) Nonlinear-response theory for steady planar couette-flow. *Phys. Rev. A*, **30**, 1528–1530.

Evans, D.J. and Morriss, G.P. (1990) *Statistical Mechanics of Nonequilibrium Liquids*, Academic Press, London.

Evans, D.J. and Rondoni, L. (2002) Comments on the entropy of nonequilibrium steady states. *J. Stat. Phys.*, **109**, 895–920.

Evans, D.J. and Searles, D.J. (2002) The fluctuation theorem. *Adv. Phys.*, **51**, 1529–1585.

Evans, D.J., Searles, D.J., and Williams, S.R. (2009) Dissipation and the relaxation to equilibrium. *J. Stat. Mech: Theory Exp.*, P07029.

Evans, D.J., Searles, D.J., and Williams, S.R. (2010) Musings on thermostats. *J. Chem. Phys.*, **133**, 104106.

Evans, D.J., Williams, S.R., and Searles, D.J. (2011) On the entropy of relaxing deterministic systems. *J. Chem. Phys.*, **135**, 194107.

Frederickson, P., Kaplan, J.L., and Yorke, J.A. (1983) The liapunov dimension of strange attractors. *J. Differ. Equ.*, **49**, 185.

Gallavotti, G. and Cohen, E.G.D. (1995a) Dynamical ensembles in nonequilibrium statistical mechanics. *Phys. Rev. Lett.*, **74**, 2694–2697.

Gallavotti, G. and Cohen, E.G.D. (1995b) Dynamical ensembles in stationary states. *J. Stat. Phys.*, **80**, 931–970.

Gaspard, P. and Nicolis, G. (1990) Transport-properties, Lyapunov exponents, and entropy per unit time. *Phys. Rev. Lett.*, **65**, 1693–1696.

Gibbs, J.W. (1981) *Elementary Principles in Statistical Mechanics*, Ox Bow Press, Woodbridge, CT.

Goldhirsch, I., Sulem, P.L., and Orszag, S.A. (1987) Stability and Lyapunov stability of dynamical systems: a differential approach and a numerical method. *Physica D*, **27**, 311–337.

Hoover, W.G. (1985) Canonical dynamics: equilibrium phase-space distributions. *Phys. Rev. A*, **31**, 1695–1697.

Hoover, W.G., Ladd, A.J.C., and Moran, B. (1982) High-strain-rate plastic flow studied Via nonequilibrium molecular dynamics. *Phys. Rev. Lett.*, **48**, 1818.

Hoover, W.G. and Posch, H.A. (1985) Direct measurement of Lyapunov exponents. *Phys. Lett. A*, **113**, 82–84.

Irving, J.H. and Kirkwood, J.G. (1950) The statistical mechanical theory of transport processes. Iv. The equations of hydrodynamics. *J. Chem. Phys.*, **18**, 817.

Kaplan, J.L. and Yorke, J.A. (1979) in *Functional Differential Equations and Approximation of Fixed Points* (eds H.O. Peitgen and H.O. Walther), Springer, Heidelberg.

Liouville, J. (1838) Note Sur La théorie De La variation Des constantes arbitraires. *J. Math.*, **3**, 349.

Maxwell, J.C. (1879) On Boltzmann's theorem on the average distribution of energy in a system of material points. *Cambridge Philos. Soc. Trans.*, **12**, 547.

Petravic, J. (2005) Time dependence of phase variables in a steady shear flow algorithm. *Phys. Rev. E*, **71**, 011202.

Ruelle, D. (1979) Ergodic theory of differentiable dynamic systems. *IHES Publ. Math.*, **50**, 27–58.

Weeks, J.D., Chandler, D., and Andersen, H.C. (1971) Role of repulsive forces in determining the equilibrium structure of simple liquids. *J. Chem. Phys.*, **54**, 5237–5247.

3
The Evans–Searles Fluctuation Theorem

> Hence the Second Law of thermodynamics is continually being violated and that to a considerable extent in any sufficiently small group of molecules belonging to any real body. As *the number of molecules in the group is increased,* the deviations from the mean of the whole become smaller and less frequent; and when the number is increased till the group includes a sensible portion of the body, the probability of a measurable variation from the mean occurring *in a finite number of years* becomes so small that it may be regarded as practically an impossibility.
>
> *(Maxwell, 1878, p. 280)*

3.1
The Transient Fluctuation Theorem

The first proof (1994) of any fluctuation theorem was for a special case of what is now known as the Evans–Searles transient fluctuation theorem (ESFT). Here we give a very general proof. Consider the response of a system, initially in some known but arbitrary distribution,

$$f(\mathbf{\Gamma};0) = \frac{\exp[-F(\mathbf{\Gamma})]}{\int_D d\mathbf{\Gamma} \exp[-F(\mathbf{\Gamma})]} \tag{3.1}$$

where $F(\mathbf{\Gamma})$ is some arbitrary single-valued real function for which $f(\mathbf{\Gamma};0) = f(M^T\mathbf{\Gamma};0)$ (i.e., the initial distribution is an even function of the momenta), over some specified phase space domain D. $\mathbf{\Gamma}$ is the extended phase space vector, which includes the phase space vector and may include additional dynamical variables such as the volume (in, say, isobaric systems) or the thermostat multiplier associated with a possible Nosé–Hoover thermostat.

Consider any system whose dynamics is described by continuous, deterministic, time-reversible equations of motion. The equations of motion may have an applied dissipative field, or the field may be zero. If the field is zero, then in order to see anything interesting, the initial distribution should not be preserved by the equations of motion (if it is preserved, then the ESFT is completely trivial). On the

Fundamentals of Classical Statistical Thermodynamics: Dissipation, Relaxation and Fluctuation Theorems, First Edition. Denis J. Evans, Debra J. Searles, and Stephen R. Williams.

other hand, if a dissipative field is applied, then it is frequently useful to consider the case where the initial distribution is the equilibrium distribution for the field free dynamics.

We assume that the unthermostatted equations of motion satisfy the AI$\boldsymbol{\Gamma}$ condition. A thermostat may be added (e.g., as in Eq. (2.14)), but again this is not absolutely essential. The equations of motion *must*, however, be time-reversal-symmetric.

Definition

The *time-averaged dissipation* $\overline{\Omega}_t(\boldsymbol{\Gamma})$ along a trajectory originating at phase $\boldsymbol{\Gamma}$ and averaged for a time t is defined as (Evans and Searles, 2002; Searles and Evans, 2000)

$$\int_0^t ds\,\Omega(S^s\boldsymbol{\Gamma}) \equiv \ln\left(\frac{f(\boldsymbol{\Gamma};0)}{f(M^T S^t\boldsymbol{\Gamma};0)}\right) - \int_0^t \Lambda(S^s\boldsymbol{\Gamma})ds$$
$$\equiv \overline{\Omega}_t(\boldsymbol{\Gamma})t \equiv \Omega_t(\boldsymbol{\Gamma}) \tag{3.2}$$

It is useful to define $\boldsymbol{\Gamma}^* \equiv M^T S^t\boldsymbol{\Gamma}$. From Eq. (2.9) we know that this phase space vector is the origin of the conjugate antitrajectory to that trajectory starting at $\boldsymbol{\Gamma}$. Going forward in time with the natural propagator from $\boldsymbol{\Gamma}^*$ is like going backwards in time from $S^t\boldsymbol{\Gamma}$ except that the velocities are reversed – see Eq. (2.9).

Definition

A system is said to be *ergodically consistent* over a phase space domain D if

$$\forall\boldsymbol{\Gamma} \in D, \quad st\ f(\boldsymbol{\Gamma};0) \neq 0,$$
$$M^T S^t\boldsymbol{\Gamma} \in D, \quad \text{and} \quad f(M^T S^t\boldsymbol{\Gamma};0) \neq 0, \ \ \forall t \tag{3.3}$$

In order for the dissipation function to be well defined over the phase space domain D, the system must be *ergodically consistent* over D. There are systems that fail to satisfy this condition. For example, if we let the initial distribution be microcanonical, and if the dynamics does not preserve the energy (there may be a dissipative field but no ergostat, etc.), then ergodic consistency obviously breaks down.

Ergodic consistency also implies that, for almost all trajectories that start at a phase vector $\boldsymbol{\Gamma}$ inside the domain D, the conjugate antitrajectory that starts at $M^T S^t\boldsymbol{\Gamma}$ is also inside D. We say "almost all" because, if there is a zero measure set of trajectories that have missing antitrajectories, this will not violate Eq. (3.3). Ergodic consistency is concerned with phase space density but not with zero measure objects (e.g., individual phase space trajectories). As mentioned in Chapter 1, almost all Loschmidt's antitrajectories exist in the initial distribution of states.

We note that the phase space domain should be specified in consideration of phase space averages, although this is often not done. If N particles are physically constrained to be located in a physical region (by impenetrable walls or so), then the specification of the domain can be very important.

We can rewrite the definition of the dissipation function so that it directly gives the ratio of the probabilities p at time zero, of observing sets of phase space trajectories originating inside infinitesimal volumes of phase space δV_Γ and $\delta V_\Gamma(\boldsymbol{\Gamma}^*) \equiv \delta V_\Gamma(M^T S^t \boldsymbol{\Gamma})$.

$$
\begin{aligned}
\frac{p(\delta V_\Gamma(\boldsymbol{\Gamma};0))}{p(\delta V_\Gamma(\boldsymbol{\Gamma}^*;0))} &= \frac{f(\boldsymbol{\Gamma};0)\delta V_\Gamma(\boldsymbol{\Gamma})}{f(\boldsymbol{\Gamma}^*;0)\delta V_\Gamma(\boldsymbol{\Gamma}^*)} \\
&\overset{\lim \delta V_\Gamma \to 0}{\to} \frac{f(\boldsymbol{\Gamma};0)}{f(\boldsymbol{\Gamma}^*;0)} \exp\left[-\int_0^t ds\ \Lambda\,(S^s\boldsymbol{\Gamma})\right] \\
&= \exp[\overline{\Omega}_t(\boldsymbol{\Gamma})t]
\end{aligned}
\tag{3.4}
$$

We have used Eq. (2.36) for $\delta V_\Gamma(\boldsymbol{\Gamma})/\delta V_\Gamma(S^t\boldsymbol{\Gamma})$, together with the observation that the Jacobian for the time reversal map is unity, $\delta V_\Gamma(M^T\boldsymbol{\Gamma}^*)/\delta V_\Gamma(\boldsymbol{\Gamma}^*) = 1$. The third line follows by a trivial use of the definition of Eq. (3.2).

Throughout this book, we will assume that the initial distribution is invariant under the time-reversal mapping, M^T and $f(M^T\boldsymbol{\Gamma};0) = f(\boldsymbol{\Gamma};0)$.

Eq. (3.4) shows that the time integral of the dissipation function gives the logarithm of the probability ratio of observing, at time zero, an infinitesimal set of trajectories relative to the conjugate set of anti-trajectories. Thus one way to think of the dissipation function is as a measure of the temporal asymmetry inherent in sets of trajectories originating from an initial distribution of states. As noted in Evans, Williams, and Searles (2010), its ensemble average is the relative entropy or Kullback–Leibler divergence (see also Kawai, Parrondo, and Broeck (2007)). As we will see, the dissipation function has an extensive set of properties.

What is not so obvious is that this definition of the time-averaged dissipation function even applies to some non-autonomous systems. If a time-dependent external field has a definite parity under time reversal over some given interval of time $[0, t]$, the conjugate sets of trajectories and antitrajectories still exist and the time-averaged dissipation can still be calculated using Eq. (3.4).

We have not said anything about how we could choose δV_Γ. Now suppose we choose the volume element δV_Γ to be the set of volume elements in D within which all trajectories originating at time zero from within that volume have the time-averaged dissipation function $\overline{\Omega}_t(\boldsymbol{\Gamma}) = (A \pm \delta A)$. Then we have

$$
\begin{aligned}
\lim_{\delta A \to 0} \frac{p(\delta V_\Gamma(\boldsymbol{\Gamma};0))}{p(\delta V_\Gamma(\boldsymbol{\Gamma}^*;0))} &= \lim_{\delta A \to 0} \frac{f(\boldsymbol{\Gamma};0)\delta V_\Gamma(\boldsymbol{\Gamma})}{f(\boldsymbol{\Gamma}^*;0)\delta V_\Gamma(\boldsymbol{\Gamma}^*)} \\
&= \exp[\overline{\Omega}_t(\boldsymbol{\Gamma})t] \\
&= \exp[At]
\end{aligned}
\tag{3.5}
$$

Using the time reversal symmetry of the equations of motion, all trajectories originating within δV_{Γ^*} must have the property $\overline{\Omega}_t(\boldsymbol{\Gamma}^*) = -(A \pm dA)$ and therefore we see that

$$
\frac{p(\overline{\Omega}_t = A)}{p(\overline{\Omega}_t = -A)} = \exp[At] \tag{3.6}
$$

This is the ESFT.

The result is clearly asymmetric. The integrated dissipation function itself is odd under time reversal. In Eq. (3.6), if A is positive, then Eq. (3.6) says it is exponentially more likely to observe positive rather than negative time-averaged dissipation. If, on the other hand, A is negative, then it is exponentially more *unlikely* to observe negative rather than positive time-averaged dissipation. Regardless of the sign of A, the implication is the same: positive time-averaged dissipation is more likely than its complementary negative counterpart.

What is not so obvious is that, for a given system, there may be multiple *noncontiguous* phase space subvolumes each of which has a time-averaged dissipation equal to $A \pm dA$. However, because the system is ergodically consistent, every such subvolume has its own conjugate phase space subvolume that contains the phase space vectors for the time-reversed conjugate antitrajectories. Every such subvolume has a time-averaged dissipation of $-A \pm dA$. Equation (3.6) is still valid in this case because, for each conjugate set of trajectories, the ratio on the left-hand side of Eq. (3.6) is still $\exp[At]$, so summing over all the noncontiguous domains leaves the ratio unchanged.

We need to stress again the conditions required for the derivation of Eq. (3.6):

- The initial distribution should be an even function of the momenta.
- The system is ergodically consistent over the relevant domain.
- The dynamics must be time-reversal-symmetric.
- The dynamics should be smooth.
- Any time-dependent external fields must have a definite parity under time reversal, over the given time interval.

Some of the conditions can be relaxed for certain systems, however we will not discuss these in this book.

Since the time-integrated dissipation function itself is extensive in the integration time and in the number of degrees of freedom, we see that for macroscopic systems observed for macroscopic times the probability of observing negative time-integrated dissipation "becomes so small that it may be regarded as practically an impossibility" – (Maxwell, 1878). It is interesting to note that Maxwell recognized the importance of *both* time and system size in relation to observing violations of the second "Law." The quote reveals that Maxwell would not be surprised at the qualitative implications of the ESFT. However, the ESFT gives a precise quantification of the matter.

It should be noted that the ESFT gives a relation between probabilities of time integrals of the dissipation function. These time integrals start at the time when the dissipation function is defined (Eq. (3.2)). The dissipation function is a functional of *both* the dynamical equations of motion that determine $S^t\mathbf{\Gamma} = \exp[iL(\mathbf{\Gamma})t]\mathbf{\Gamma}$ from the initial phase $\mathbf{\Gamma}$ and also the initial distribution $f(\mathbf{\Gamma}; 0)$. These "initial" times need to be one and the same.

The instantaneous dissipation function can be determined by differentiation of Eq. (3.2):

$$
\begin{aligned}
\frac{\partial}{\partial t}\int_0^t ds\,\Omega(S^s\mathbf{\Gamma}) &= \Omega(S^t\mathbf{\Gamma})\\
&= \frac{\partial}{\partial t}\left[\ln\left(f(\mathbf{\Gamma};0)\right) - \ln\left(f(e^{iL(\mathbf{\Gamma})t}\mathbf{\Gamma};0)\right) - \int_0^t ds\,\Lambda(e^{iL(\mathbf{\Gamma})s}\mathbf{\Gamma})\right]\\
&= -\frac{1}{f(e^{iL(\mathbf{\Gamma})t}\mathbf{\Gamma};0)}\frac{\partial f(e^{iL(\mathbf{\Gamma})t}\mathbf{\Gamma};0)}{\partial t} - \Lambda(e^{iL(\mathbf{\Gamma})t}\mathbf{\Gamma})\\
&= -\frac{1}{f(e^{iL(\mathbf{\Gamma})t}\mathbf{\Gamma};0)}\frac{\partial e^{iL(\mathbf{\Gamma})t}\mathbf{\Gamma}}{\partial t}\cdot\frac{\partial f(e^{iL(\mathbf{\Gamma})t}\mathbf{\Gamma};0)}{\partial(e^{iL(\mathbf{\Gamma})t}\mathbf{\Gamma})} - \Lambda(e^{iL(\mathbf{\Gamma})t}\mathbf{\Gamma})\\
&= -\frac{1}{f(e^{iL(\mathbf{\Gamma})t}\mathbf{\Gamma};0)}[iL(\mathbf{\Gamma})e^{iL(\mathbf{\Gamma})t}\mathbf{\Gamma}]\cdot\frac{\partial f(e^{iL(\mathbf{\Gamma})t}\mathbf{\Gamma};0)}{\partial(e^{iL(\mathbf{\Gamma})t}\mathbf{\Gamma})} - \Lambda(e^{iL(\mathbf{\Gamma})t}\mathbf{\Gamma})\\
&= -\frac{1}{f(S^t\mathbf{\Gamma};0)}\dot{\mathbf{\Gamma}}(S^t\mathbf{\Gamma})\cdot\frac{\partial f(S^t\mathbf{\Gamma};0)}{\partial(S^t\mathbf{\Gamma})} - \Lambda(S^t\mathbf{\Gamma}). \qquad (3.7)
\end{aligned}
$$

The derivative on the left-hand side of Eq. (3.7) is to be computed at a fixed point in phase space. It we now set $t = 0$, we obtain an expression for the instantaneous dissipation function:

$$
\Omega(\mathbf{\Gamma}) = -\frac{1}{f(\mathbf{\Gamma};0)}\dot{\mathbf{\Gamma}}(\mathbf{\Gamma})\cdot\frac{\partial f(\mathbf{\Gamma};0)}{\partial\mathbf{\Gamma}} - \Lambda(\mathbf{\Gamma}) \qquad (3.8)
$$

The ESFT has generated much interest, as it shows how irreversibility emerges from the deterministic, time-reversible equations of motion. Its proof is extremely simple and uses almost nothing but the time reversibility of the underlying dynamics. Because its proof relies on so few assumptions, the ESFT is extremely general. It is valid arbitrarily far from equilibrium. It applies to systems of arbitrary size. Taking the classical "thermodynamic limit" is *not* required.

It provides a generalized form of the second "Law" of thermodynamics that can be applied to small systems observed for short periods. It also resolves the long-standing Loschmidt Paradox. The ESFT has been verified experimentally: for examples, see Wang *et al.* (2002), Carberry *et al.* (2004a, 2007), Reid *et al.* (2004), Collin *et al.* (2005), Liphardt *et al.* (2002), Trepagnier *et al.* (2004), Schuler *et al.* (2005).

The *form* of the ESFT Eq. (3.6) applies to any valid ensemble/dynamics combination. However, the precise *expression* for $\overline{\Omega}_t$ given in Eq. (3.2) is dependent on both the initial distribution *and* the dynamics.

3.2 Second Law Inequality

We are now in a position to use the ESFT to derive a number of simple inequalities. The derivation of the second law inequality (SLI) from the ESFT provides what amounts to a proof of the second "Law" of thermodynamics. The SLI shows that *time averages* (rather than instantaneous values) of the ensemble-averaged dissipation are nonnegative. This SLI is valid for the appropriately time-averaged dissipation but the ensemble-averaged instantaneous dissipation may be negative for intermediate times.

The SLI states that (Searles and Evans, 2004)

$$\langle \overline{\Omega}_t \rangle \geq 0, \quad \forall t > 0 \tag{3.9}$$

The proof is trivial and is obtained by integration of Eq. (3.6):

$$\begin{aligned}
\langle \overline{\Omega}_t \rangle &= \int_{-\infty}^{+\infty} dB\, p(\overline{\Omega}_t = B)B \\
&= \int_{0}^{+\infty} dB\, p(\overline{\Omega}_t = B)B + \int_{-\infty}^{0} dB\, p(\overline{\Omega}_t = B)B \\
&= \int_{0}^{+\infty} dB\, p(\overline{\Omega}_t = B)B - \int_{0}^{+\infty} dB\, p(\overline{\Omega}_t = -B)B \\
&= \int_{0}^{+\infty} dB\, p(\overline{\Omega}_t = B)B(1 - \exp[-B]) \geq 0
\end{aligned} \tag{3.10}$$

In linear, irreversible thermodynamics, it is asserted that the quantity called the *spontaneous entropy* production cannot be negative. Close to equilibrium, the ensemble-averaged dissipation for a driven system is equal to the ensemble-averaged spontaneous entropy production (see Chapter 6). In an electric circuit close to equilibrium, both quantities are equal to the product of the electric current times the voltage divided by the ambient temperature. If the circuit has a complex impedance, there will necessarily be a phase lag between the applied voltage and the current. This means that for an AC sinusoidal electric circuit, there will always be intervals within a cycle within which the dissipation or entropy production is negative. This presents serious difficulties for linear, irreversible thermodynamics, but the SLI is not presented with any difficulties by this matter. The SLI only asserts that the time-integrated, ensemble-averaged dissipation is positive. The time integral begins at the initial time when the dissipation function itself was defined. The SLI does *not* state that the ensemble-averaged instantaneous dissipation must be positive.

Now let us look at Eq. (3.10) in more detail. For *every* value of $B > 0$ if $p(\overline{\Omega}_t = B) > 0$, ergodic consistency implies $p(\overline{\Omega}_t = -B) > 0$. This is because for every set of trajectories the conjugate set of antitrajectories exists within the ostensible phase space domain. Furthermore, as can be seen from Eq. (3.10), the fluctuation theorem shows $p(\overline{\Omega}_t = B) > p(\overline{\Omega}_t = -B) \neq 0$. This, in turn, means that if the time-integrated dissipation is nonzero for some infinitesimal set of initial points in phase space near $\mathbf{\Gamma}$

$$\overline{\Omega}_t(\mathbf{\Gamma}) \neq 0, \quad \mathbf{\Gamma} \in D, \quad t > 0 \Rightarrow \langle \overline{\Omega}_t \rangle > 0 \tag{3.11}$$

Definition

A nonzero value for the time-averaged dissipation of an infinitesimal set of phase space trajectories anywhere in the ostensible phase space domain implies, for ergodically consistent systems, a *strict SLI* Eq. (3.11).

We note that the SLI (both strict and otherwise) has *macroscopic* consequences for the fluctuation theorem. The SLI has important consequences in widely varied applications such as atmospheric physics and aerodynamics.

3.3 Nonequilibrium Partition Identity

This Identity (also referred to as the *Kawasaki identity, Kawasaki normalization factor, Kawasaki function*, and *the integral fluctuation theorem*) was first implied for Hamiltonian systems by Yamada and Kawasaki in 1967 and was explicitly noted by Morriss and Evans for thermostatted systems driven by an external field in 1984 (Morriss and Evans, 1985; Evans and Searles, 1995; Carberry *et al.*, 2004b). The nonequilibrium partition identity (NPI) is stated as follows:

$$\left\langle \exp\left[-\overline{\Omega}_t t\right]\right\rangle = 1 \tag{3.12}$$

A very simple proof can be obtained using the ESFT given in Eq. (3.6):

$$\begin{aligned}\left\langle \exp\left[-\overline{\Omega}_t t\right]\right\rangle &= \int_{-\infty}^{+\infty} dA\, p(\overline{\Omega}_t = A)\exp[-At]\\ &= \int_{-\infty}^{+\infty} dA\, p(\overline{\Omega}_t = -A)\\ &= \int_{-\infty}^{+\infty} dA'\, p(\overline{\Omega}_t = A') = 1\end{aligned} \tag{3.13}$$

It is, at first, quite extraordinary that, although the SLI says the exponent of the NPI is negative on average, the rare instances when the dissipation function has a negative time average occur with such frequency that their exponentially enhanced effect ensures the average of the exponential is always unity. Trivially, we observe that the NPI is still valid even in the case where $\Omega_t(\mathbf{\Gamma}) = 0,\ \forall \mathbf{\Gamma} \in D$.

We note that, in order to *observe* the NPI in real experimental data, we must be able to observe the antitrajectories that are conjugate to the most probable trajectories. In macroscopic systems, this may be (as Maxwell already noted) impossible because of the extremely low probability of observing these events.

For real data, one can only expect to observe time-averaged values of the dissipation over some finite range. In some experiments, no negative dissipation averages may be observed. In this case, the NPI cannot be experimentally verified. Even when negative values are observed, they will generally have a more restricted range than for the averages that are positive. In such cases, one can *prune* the distribution so that it is bounded $-B \leq \overline{\Omega}_t \leq +B$ with $\widetilde{p}(\overline{\Omega}_t = b) \neq 0,\ \forall |b| \leq B$ for a bounded, normalized probability distribution $\widetilde{p}$ in order to obtain an experimentally verifiable result. Then we can write

$$\begin{aligned}&\left\langle \exp\left[-\overline{\Omega}_t t\right]\right\rangle_{-B\leq\overline{\Omega}_t\leq+B}\\ &= \int_{0}^{+B} dA\,\widetilde{p}(\overline{\Omega}_t = A)\exp[-At] + \int_{-B}^{0} dA\,\widetilde{p}(\overline{\Omega}_t = A)\exp[-At]\end{aligned}$$

$$
\begin{aligned}
&= \int_{0}^{+B} dA\,\tilde{p}(\overline{\Omega}_t = -A) + \int_{-B}^{0} dA\,\tilde{p}(\overline{\Omega}_t = -A) \\
&= \int_{-B}^{B} dA\,\tilde{p}(\overline{\Omega}_t = -A) \\
&= \int_{-B}^{B} dA'\,\tilde{p}(\overline{\Omega}_t = A') = 1 \\
&= \int_{-B}^{+B} dA'\,\tilde{p}(\overline{\Omega}_t = A't) = 1
\end{aligned}
\qquad (3.14)
$$

This restricted range distribution has a vastly better behaved average than the corresponding unrestricted average in Eq. (3.13). In Eq. (3.13), the average almost always approaches unity from below, making it extremely difficult to estimate numerical uncertainties in actual experimental data.

Pruning the probability distribution guarantees ergodic consistency in the empirical data (i.e., if a value of $\overline{\Omega}_t = A \pm \delta A$ is observed, then so too is $\overline{\Omega}_t = -A \pm \delta A$ $-\mathrm{A} \pm \mathrm{dA}$). The unpruned distribution violates ergodic consistency of the empirical data. One can observe sets of trajectories with positive time-averaged dissipation but not observe any of their conjugate antitrajectories. Increasing the sample size widens the observed range of time-averaged dissipation but the unpruned distribution will always be ergodically inconsistent sufficiently far from zero for the negative dissipation states. Furthermore, if the numerical error in the experimentally determined $p(\overline{\Omega}_t = -A)$ is very large, then it is it is better to exclude it from the integral in the evaluation of Eq. (3.13), as its magnitude will be amplified by e^{At} in that expression. This is achieved in Eq. (3.14) by restricting the bounds.

Lastly, we note that, although the ESFT implies the NPI, the converse is not true (Carberry *et al.*, 2004b).

3.4
Integrated Fluctuation Theorem (Evans and Searles, 2002)

The fluctuation theorem quantifies the probability of observing time-averaged dissipation functions having complimentary values. The SLI only states that the ensemble average of the time-averaged dissipation should be positive rather than negative. Therefore, it is of interest to construct a fluctuation theorem that predicts the probability ratio that the time-averaged dissipation function is positive rather than negative.

In experimental situations where the statistical error is large and the ensemble sample sizes are small, it is useful to be able to predict the probability that the time-averaged dissipation is negative. The integrated form of the fluctuation theorem (IFT) (Ayton, Evans, and Searles, 2001) gives a relationship that quantifies the probability of observing Second Law violations in small systems observed for a short time.

The ESFT, Eq. (3.6), can be written as

$$\frac{p(\overline{\Omega}_t = -A)}{p(\overline{\Omega}_t = A)} = \exp(-At) \tag{3.15}$$

We wish to give the probability ratio of observing trajectories with positive and negative values of $\overline{\Omega}_t$, and so we consider

$$p_+(t) \equiv p(\overline{\Omega}_t > 0), \quad p_-(t) \equiv p(\overline{\Omega}_t < 0) \tag{3.16}$$

Now

$$\frac{p_-(t)}{p_+(t)} = \frac{\int_0^\infty dA\, p(\overline{\Omega}_t = -A)}{\int_0^\infty dA\, p(\overline{\Omega}_t = A)} \tag{3.17}$$

Using Eq. (3.6)

$$\frac{p_-(t)}{p_+(t)} = \frac{\int_0^\infty dA \exp(-At) p(\overline{\Omega}_t = A)}{\int_0^\infty dA\, p(\overline{\Omega}_t = A)} \tag{3.18}$$

The right-hand side of this equation is just the ensemble average of $\exp(-\overline{\Omega}_t t)$ evaluated over that subset of trajectories for which the time-averaged dissipation is positive.

Again, if we look at Eqs. (3.17) and (3.18) in detail, we see that on the right-hand side for *every* value of $A > 0$ $p(\overline{\Omega}_t = -A) < p(\overline{\Omega}_t = A)$. This, in turn, means that, if the time-integrated dissipation is nonzero for *any* value of A

$$p(\overline{\Omega}_t(\mathbf{\Gamma}) = A) \neq 0, \quad \Rightarrow \frac{p_-(t)}{p_+(t)} = \langle \exp(-\overline{\Omega}_t t) \rangle_{\overline{\Omega}_t > 0} < 1 \tag{3.19}$$

From Eq. (3.18) we can also obtain the reciprocal relationship:

$$\frac{p_+(t)}{p_-(t)} = \frac{1}{\langle \exp(-\overline{\Omega}_t t) \rangle_{\overline{\Omega}_t > 0}} \geq 1 \tag{3.20}$$

where the equality holds only if $p(\Omega_t = A) = 0, \ \forall A$.

Similarly, it can be shown that

$$\frac{p_+(t)}{p_-(t)} = \langle \exp(-\overline{\Omega}_t t) \rangle_{\overline{\Omega}_t < 0} \geq 1 \tag{3.21}$$

where the equality holds only if $p(\Omega_t = A) = 0, \ \forall A$.

We note that in actual experiments where $\langle \overline{\Omega}_t \rangle > 0$, Eqs. (3.19) and (3.20) have much smaller statistical uncertainties than Eq. (3.21), because rarely observed trajectory segments with highly negative values of $\overline{\Omega}_t$ will have a large influence on the ensemble average in Eq. (3.21). Consequently, Eq. (3.21) should be avoided in numerical calculations or laboratory experiments.

Finally, we note that Eq. (3.21) can be used to show that

$$p_-(t) = \frac{\left\langle \exp\left(-\overline{\Omega}_t t\right)\right\rangle_{\overline{\Omega}_t>0}}{\left(1+\left\langle \exp\left(-\overline{\Omega}_t t\right)\right\rangle_{\overline{\Omega}_t>0}\right)}, \quad p_+(t) = \frac{1}{\left(1+\left\langle \exp\left(-\overline{\Omega}_t t\right)\right\rangle_{\overline{\Omega}_t>0}\right)} \tag{3.22}$$

Obviously, $p_-(t) + p_+(t) = 1, \ \forall t$ and, again, $p_-(t) \leq p_+(t), \ \forall t$. Furthermore, one can have equality only if there is zero dissipation. Nonzero dissipation anywhere in the relevant phase space implies $p_-(t) < p_+(t), \ \forall t$!

3.5
Functional Transient Fluctuation Theorem (Evans and Searles, 2002)

The FTs derived above predict the ratio of the probabilities of observing conjugate values of the dissipation function. As given above, these theorems give no information on the probability ratios for any functions other than the dissipation function, Eq. (3.2). In this section we describe how the FT can be extended to apply to arbitrary phase functions that have an odd parity under time reversal (Searles, Ayton, and Evans, 2000).

Let $\phi(\boldsymbol{\Gamma})$ be an arbitrary phase function, and define the time average

$$\overline{\phi}_{i,t} = \frac{1}{t}\int_0^t ds\,\phi(S^s\boldsymbol{\Gamma}_i) \tag{3.23}$$

for a phase space trajectory: $S^s\boldsymbol{\Gamma}_i;\ 0 < s < t$. At $t=0$, the phase space volume occupied by a contiguous set of trajectories for which $\{\boldsymbol{\Gamma}_i | A - \delta A < \overline{\phi}_{i,t} < A + \delta A\}$ is given by $\delta V_{\boldsymbol{\Gamma}}(\boldsymbol{\Gamma})$, and at time t these phase points will occupy a volume $\delta V_{\boldsymbol{\Gamma}}(S^t\boldsymbol{\Gamma}) = \delta V_{\boldsymbol{\Gamma}}(\boldsymbol{\Gamma})e^{\overline{\Lambda}_t t}$, where $\overline{\Lambda}_t$ is the time-averaged phase space expansion factor along these trajectories – see Eq. (2.36). We denote $\overline{\phi}(t) = \langle \overline{\phi}_{i,t}\rangle_{\{i\}}$, that is, the average value of $\overline{\phi}_{i,t}$ over the set of contiguous trajectories $\{\boldsymbol{\Gamma}_i\}$.

If the dynamics is reversible and the system is ergodically consistent, there will be a contiguous set of initial phases $\{\boldsymbol{\Gamma}_i^*\}$, given by $\boldsymbol{\Gamma}_i^* = M^T(S^t\boldsymbol{\Gamma}_i)$, that will occupy a volume $\delta V_{\boldsymbol{\Gamma}}(\boldsymbol{\Gamma}^*) = \delta V_{\boldsymbol{\Gamma}}(S^t\boldsymbol{\Gamma}) = \delta V_{\boldsymbol{\Gamma}}(\boldsymbol{\Gamma})e^{\overline{\Lambda}_t t}$ along which the time-averaged value of the phase function is $\overline{\phi}_{i*,t} = M^T(\overline{\phi}_{i,t})$. For any $\phi_i(\boldsymbol{\Gamma})$ that is odd under time reversal, $\overline{\phi}_{i*,t} = -\overline{\phi}_{i,t}$.

The probability ratio of observing trajectories originating in an initial phase volume and its conjugate phase volume will be related to the initial phase space distribution function and the measure of the volume elements by Eq. (3.4). Therefore, from the definition of the dissipation function in Eq. (3.2) we obtain

$$\lim_{\delta V_{\boldsymbol{\Gamma}}\to 0}\frac{p(\delta V_{\boldsymbol{\Gamma}}(\boldsymbol{\Gamma};0))}{p(\delta V_{\boldsymbol{\Gamma}}(\boldsymbol{\Gamma}^*;0))} = \exp[\overline{\Omega}_t(\boldsymbol{\Gamma})t] \tag{3.24}$$

If the phase function is odd under time reversal symmetry, then the ratio of the probability of observing trajectories for which $A - \delta A < \overline{\phi}_t < A + \delta A$ to the probability of observing conjugate trajectories, for which $-A - \delta A < \overline{\phi}_t < -A + \delta A$, is

$$\frac{p(\overline{\phi}_t = A)}{p(\overline{\phi}_t = -A)} = \frac{\displaystyle\int_{\overline{\phi}_t(\mathbf{\Gamma})=A} d\mathbf{\Gamma} f(\mathbf{\Gamma};0)}{\displaystyle\int_{\overline{\phi}_t(\mathbf{\Gamma}^*)=-A} d\mathbf{\Gamma}^* f(\mathbf{\Gamma}^*;0)} = \frac{\displaystyle\int_{\overline{\phi}_t(\mathbf{\Gamma})=A} d\mathbf{\Gamma} f(\mathbf{\Gamma};0)}{\displaystyle\int_{\overline{\phi}_t(\mathbf{\Gamma})=A} d\mathbf{\Gamma} f(\mathbf{\Gamma};0) e^{-\overline{\Omega}_t(\mathbf{\Gamma})t}} = \left\langle e^{-\overline{\Omega}_t t} \right\rangle_{\overline{\phi}_t=A}^{-1} \tag{3.25}$$

where the notation $\langle \cdots \rangle_{\overline{\phi}_t=A}$ refers to the *ensemble average* over (possibly) noncontiguous trajectory sets for which $A - \delta A < \overline{\phi}_t < A + \delta A$. Equation (3.25) gives the ratio of the measure of those phase space trajectories for which $\overline{\phi}_t = A$ to the measure of those trajectories for which $\overline{\phi}_t = -A$.

This is the functional transient fluctuation theorem (FTFT) for any phase variable $\overline{\phi}_t$ that is odd under time reversal. Provided it has a definite parity under time reversal symmetry, the actual form of $\overline{\phi}_t$ is quite arbitrary.

If the phase variable is even under time reversal symmetry, then we cannot obtain the relationship between $p(\overline{\phi}_t = A)/p(\overline{\phi}_t = -A)$ using the time reversal properties of the dissipation function.

3.6 The Covariant Dissipation Function

As we have seen already, the dissipation function is a rather important function in statistical mechanics. In later chapters we will see that it plays a key role in almost all aspects of nonequilibrium statistical mechanics – in response theory and in understanding the process of relaxation toward equilibrium. It is defined in terms of the initial distribution of states and also by the dynamical equations of motion.

What happens to the dissipation function if we redefine the dissipation function in terms of the time-evolving N-particle phase space distribution function rather than the initial distribution? The time-covariant dissipation function could be written as

$$\Omega_\tau(S^{t_1}\mathbf{\Gamma}; t_1) \equiv \ln\left(\frac{f\left(S^{t_1}\mathbf{\Gamma}; t_1\right)}{f(M^T S^{t_1+\tau}\mathbf{\Gamma}; t_1)}\right) - \int_{t_1}^{t_1+\tau} \Lambda(S^s\mathbf{\Gamma}) ds \tag{3.26}$$

where the dissipation function is integrated for a time τ but defined with respect to the phase space density at time t_1 rather than at the usual time zero. By constructing the ESFT at time t_1 and allowing this time to increase without bound, we could construct an exact, steady-state FT for thermostatted driven systems that evolved toward nonequilibrium steady states. This steady-state FT would not be asymptotic, unlike the FT to be discussed in Section 6.10.

However, there is a serious problem posed by this scenario. The time-integrated dissipation function is related to a number of important physical properties. We have met only a small number of these properties thus far in this book. This means that there must be some kind of invariance of properties satisfied by the dissipation function. So you really cannot constantly redefine the quantity.

From the definition Eq. (3.26) we see that

$$\lim_{\delta V_\Gamma \to 0} \frac{p\left[\delta V_\Gamma\left(S^{t_1}\boldsymbol{\Gamma}\right);t_1\right]}{p\left[\delta V_\Gamma\left(M^T S^{t_1+\tau}\boldsymbol{\Gamma}\right);t_1\right]} = \lim_{\delta V_\Gamma \to 0} \frac{f(S^{t_1}\boldsymbol{\Gamma};t_1)\delta V_\Gamma(S^{t_1}\boldsymbol{\Gamma})}{f(M^T S^{t_1+\tau}\boldsymbol{\Gamma};t_1)\delta V_\Gamma(M^T S^{t_1+\tau}\boldsymbol{\Gamma})}$$
$$= \exp\left[\Omega_\tau\left(S^{t_1}\boldsymbol{\Gamma};t_1\right)\right] \tag{3.27}$$

Now all the trajectories that arrive in $\delta V_\Gamma(S^{t_1}\boldsymbol{\Gamma})$ at time t_1 started out within $\delta V_\Gamma(\boldsymbol{\Gamma})$ at time zero. All the trajectories that arrive at $\delta V_\Gamma(S^{t_1+\tau}\boldsymbol{\Gamma})$ at time $t_1+\tau$ would have continued on to $\delta V_\Gamma(S^{2t_1+\tau}\boldsymbol{\Gamma})$ at time $2t_1+\tau$. Furthermore, all trajectories within the volume element $\delta V_\Gamma(S^{2t_1+\tau}\boldsymbol{\Gamma})$ at time $2t_1+\tau$ started within $\delta V_\Gamma(\boldsymbol{\Gamma})$ at time zero. So in fact

$$\lim_{\delta V_\Gamma \to 0} \frac{p\left[\delta V_\Gamma\left(S^{t_1}\boldsymbol{\Gamma}\right);t_1\right]}{p\left[\delta V_\Gamma\left(M^T S^{t_1+\tau}\boldsymbol{\Gamma}\right);t_1\right]} = \lim_{\delta V_\Gamma \to 0} \frac{p[\delta V_\Gamma(\boldsymbol{\Gamma});0]}{p\left[\delta V_\Gamma\left(M^T S^{2t_1+\tau}\boldsymbol{\Gamma}\right);0\right]} \tag{3.28}$$

For the antitrajectories, going backwards in time from t_1 to zero is like going forward in time an additional amount t_1 from time $t_1+\tau$, and therefore (Evans, Searles, and Williams, 2010)

$$\Omega_\tau(S^{t_1}\boldsymbol{\Gamma};t_1) = \Omega_{2t_1+\tau}(\boldsymbol{\Gamma};0) \tag{3.29}$$

The antitrajectories at time zero to those within $\delta V_\Gamma(\boldsymbol{\Gamma})$ are the time reversal mapped phases of those in $\delta V_\Gamma(S^{2t_1+\tau}\boldsymbol{\Gamma})$.

So there is no new information contained within the time-covariant dissipation function. This leads to the following very important observation: *There is no time-local, non-asymptotic ESFT for steady states with time-reversible deterministic dynamics.*

3.7
The Definition of Equilibrium

> If a system is very weakly coupled to a heat bath at a given 'temperature,' if the coupling is indefinite or not known precisely, if the coupling has been on for a long time, and if all the 'fast' things have happened and all the 'slow' things not, the system is said to be in *thermal equilibrium.*
>
> *(Feynman, 1972)*

One of the aims of this book is to understand the true nature of thermal equilibrium. We will return to discuss the nature of equilibrium many times in this book, each time with a little more knowledge than before, until at the end of Chapter 5 we will be able to demonstrate that the definition we are about to introduce, indeed, encompasses each aspect of the qualitative notion of equilibrium given by Feynman in the quote above.

It may seem somewhat odd that we should introduce a definition of equilibrium in a chapter that is mostly devoted to discussing nonequilibrium systems. However, you cannot really understand equilibrium without first knowing how nonequilibrium systems relax toward equilibrium.

Definition

An *equilibrium system* is characterized by an N-particle phase space distribution and a dynamics for which, over the phase space domain D, the time-integrated dissipation function is identically zero:

$$\begin{aligned}\overline{\Omega}_{\mathrm{eq},t}(\mathbf{\Gamma}) = 0, \quad &\forall\mathbf{\Gamma} \in D,\ \forall t > 0 \\ &\Rightarrow \langle(\overline{\Omega}_{\mathrm{eq},t}t)\rangle = 0, \quad \forall t > 0 \\ &\Rightarrow p_{\mathrm{eq},+}(t) = p_{\mathrm{eq},-}(t), \quad \forall t > 0\end{aligned} \tag{3.30}$$

Although this is a convenient definition of equilibrium, we do not yet know whether equilibrium systems exist or whether such systems are stable. It turns out that the answer to both these questions is yes, but these answers will be given only in the next chapters.

We have already seen that the only way $\langle\Omega_t\rangle = 0$ is if the instantaneous dissipation and the time-averaged dissipation are both zero everywhere, Eq. (3.30). Consequently the ensemble-averaged, time-integrated dissipation is zero if and only if the time-integrated dissipation is zero almost everywhere in the ostensible phase space:

$$\langle\Omega_t\rangle_{\mathrm{eq}} = 0 \Leftrightarrow \Omega_t(\mathbf{\Gamma}) = 0, \quad \forall!\mathbf{\Gamma} \in D,\ \forall t > 0 \tag{3.31}$$

Definition

Equation. (3.31) is called the *second law equality.*

A number of corollaries follow immediately. From Eq. (3.30) we observe that, for equilibrium systems that are ergodically consistent over D, the probability of observing *every* infinitesimal *set* of phase space trajectories is equal to the probability of observing, at time zero, the conjugate set of antitrajectories:

$$\frac{p_{\mathrm{eq}}(\delta V_{\mathbf{\Gamma}}(\mathbf{\Gamma});0)}{p_{\mathrm{eq}}(\delta V_{\mathbf{\Gamma}}(M^T S^t\mathbf{\Gamma});0)} = 1, \quad \forall!\mathbf{\Gamma} \in D,\ \forall t \tag{3.32}$$

The equilibrium state is therefore time-reversal-symmetric.

For instance, if we compute the Lyapunov spectrum $\{\lambda_i; i = 1, \ldots, d; \mathbf{\Gamma}(0)\}$ for a system with time-reversible dynamics, for a trajectory originating at $\mathbf{\Gamma}(0)$, we know that for any *steady* system (nonequilibrium steady state or an equilibrium state) the spectra have the property that if we reverse the direction of time, the largest, most positive exponent will be -1 times the smallest, most negative exponent of the original system. If we denote the exponents of the time-reversed system as $\{\lambda_i^*; i = 1, \ldots, d; \mathbf{\Gamma}^*(0)\}$, we will have

$$\lambda_i^*(\mathbf{\Gamma}^*(0)) = -\lambda_{f-i+1}(\mathbf{\Gamma}(0)), \quad \forall!\mathbf{\Gamma}(0) \in D \tag{3.33}$$

where f is the number of nonzero Lyapunov exponents in the system.

Now, if we further assume that the system is an ergodic equilibrium system, we see that the spectrum must be independent of the initial phase $\mathbf{\Gamma}(0)$ or $\mathbf{\Gamma}^*(0)$. This means that the spectrum for the trajectory must be the same as the spectrum of the antitrajectories. At equilibrium, therefore the time reversal map transforms the spectrum into itself. This means that, at equilibrium

$$\lambda_{\mathrm{eq},i} = -\lambda_{\mathrm{eq},f-i+1}(\mathbf{\Gamma}(0)), \;\; \forall i \tag{3.34}$$

Definition

This is termed the *conjugate pairing rule for equilibrium systems.*

All ergodic equilibrium systems have Lyapunov spectra that, apart from any unpaired zero exponents, consist of conjugate pairs of exponents that each sum to zero. The conjugate paired exponents define sets of two-dimensional areas that are each preserved in measure by the natural dynamics. The Kaplan–Yorke dimension of an ergodic equilibrium system is equal to the number of Lyapunov exponents (including unpaired zero exponents), which is also the ostensible dimension of the phase space. For these systems, the ostensible phase space volume is at least, on average, preserved by the natural dynamics.

Thus far we have only discussed equilibrium systems in the context of time-integrated dissipation. Later, in Chapter 4 we will talk about equilibrium in the context of instantaneous dissipation. At the moment we do not know whether $\Omega_t(\mathbf{\Gamma}) = 0, \;\forall\mathbf{\Gamma} \in D \Rightarrow \Omega_{t+\tau}(\mathbf{\Gamma}) = 0, \;\forall\mathbf{\Gamma} \in D, t, \tau > 0$. These questions and others will be answered in the next chapter.

3.8 Conclusion

One often sees in the historical, and even in the recent, literature statements that imply irreversibility results from the special nature of the initial state. For example

> it is in any case impossible on the basis of *present* theory to carry out a mechanical derivation of the second law without specializing the initial state
>
> *E. Zermelo translated by Brush (1966, pp. 194–202).*

or

> I have called it one of the most brilliant confirmations of the mechanical view of Nature that it provides an extraordinarily good picture of the dissipation of energy, as long as one assumes that the world began in an initial

state satisfying certain conditions. I have called this state an improbable state.

A quote from: "A word from mathematics to energism"
L. Boltzmann interpretted by Broda (1983, p. 74).

The time-asymmetry comes merely from the fact that the system has been *started off* in a very special (i.e., low entropy) state.

(Penrose, 1990, p. 408)

With respect to the fluctuation theorem, the initial state need not be a state of particularly low probability. Equation (3.1) is a rather general distribution function, and the FT holds for all distributions subject to the rather mild assumptions given above. This initial state simply cannot be an equilibrium state because, if this is so, everything is time-reversal-symmetric. If there is any dissipation at all, the probability of positive time-averaged dissipation is exponentially more likely (in time and in the number of degrees of freedom) than the probability of complementary negative time-averaged dissipation Eq. (3.6).

What was never realized until the proof of the fluctuation theorem was a rather simple fact. Loschmidt's assertion (that for every trajectory there exists a conjugate antitrajectory and that summing over all such conjugate pairs implies that irreversibility is impossible) is simply wrong. One must, instead, consider not individual phase space trajectories but the probabilities of infinitesimal *sets* of trajectories having specified properties within some tolerance. It is this probability ratio that gives the dissipation function its meaning. It makes no mathematical sense to think that individual conjugate trajectory pairs somehow cancel each other out. Only when the system is at equilibrium do the probabilities of observing *sets* of trajectories and their conjugate antitrajectories become equal, giving us, for the first time, a mathematical definition of equilibrium.

References

Ayton, G., Evans, D.J., and Searles, D.J. (2001) A local fluctuation theorem. *J. Chem. Phys.*, **115**, 2033–2037.

Broda, E. (1983) *Ludwig Boltzmann*, Ox Bow Press, Woodbridge, CT.

Brush, S.G. (1966) *Irreversible Processes: Kinetic Theory*, vol. **2**, Pergamon Press, Oxford.

Carberry, D.M., Baker, M.A.B., Wang, G.M., Sevick, E.M., and Evans, D.J. (2007) An optical trap experiment to demonstrate fluctuation theorems in viscoelastic media. *J. Opt. A*, **9**, S204.

Carberry, D.M., Reid, J.C., Wang, G.M., Sevick, E.M., Searles, D.J., and Evans, D.J. (2004a) Fluctuations and irreversibility: an experimental demonstration of a second-Law-like theorem using a colloidal particle held in an optical trap. *Phys. Rev. Lett.*, **92**, 140601.

Carberry, D.M., Williams, S.R., Wang, G.M., Sevick, E.M., and Evans, D.J. (2004b) The Kawasaki identity and the fluctuation theorem. *J. Chem. Phys.*, **121**, 8179–8182.

Collin, D., Ritort, F., Jarzynski, C., Smith, S.B., Tinoco, I., and Bustamante, C. (2005) Verification of the crooks fluctuation theorem and recovery of Rna folding free energies. *Nature*, **437**, 231–234.

Evans, D.J. and Searles, D.J. (1995) Steady states, invariant measures, and response theory. *Phys. Rev. E*, **52**, 5839–5848.

Evans, D.J. and Searles, D.J. (2002) The fluctuation theorem. *Adv. Phys.*, **51**, 1529–1585.

Evans, D.J., Searles, D.J., and Williams, S.R. (2010) The covariant dissipation function for transient nonequilibrium states. *J. Chem. Phys.*, **133**, 054507.

Evans, D.J., Williams, S.R., and Searles, D.J. (2010) Thermodynamics of small systems, in *Nonlinear Dynamics of Nanosystems* (eds G. Radons, B. Rumpf, and H.G. Schuster), Wiley-VCH Verlag GmbH, Weinheim.

Feynman, R.P. (1972) *Statistical Mechanics, A Set of Lectures*, Benjamin, Reading, MA.

Kawai, R., Parrondo, J.M.R., and van den Broeck, C. (2007) Dissipation: the phase-space perspective. *Phys. Rev. Lett.*, **98**, 080602.

Liphardt, J., Dumont, S., Smith, S.B., Tinoco, I., and Bustamante, C. (2002) Equilibrium information from nonequilibrium measurements in an experimental test. *Science*, **296**, 1832–1835.

Maxwell, J.C. (1878) Tait's "Thermodynamics" Ii. *Nature*, **17**, 278–280.

Morriss, G.P. and Evans, D.J. (1985) Isothermal response theory. *Mol. Phys.*, **54**, 629–636.

Penrose, R. (1990) *The Emperor's New Mind*, Vantage, London.

Reid, J.C., Carberry, D.M., Wang, G.M., Sevick, E.M., Evans, D.J., and Searles, D.J. (2004) Reversibility in nonequilibrium trajectories of an optically trapped particle. *Phys. Rev. E*, **70**, 016111/1–016111/9.

Schuler, S., Speck, T., Tietz, C., Wrachtrup, J., and Seifert, U. (2005) Experimental test of the fluctuation theorem for a driven Two-level system with time-dependent rates. *Phys. Rev. Lett.*, **94**, 180602.

Searles, D.J., Ayton, G.S., and Evans, D.J. (2000) Generalised fluctuation formula. *AIP Conf. Proc.*, **159**, 271–280.

Searles, D.J. and Evans, D.J. (2000) Ensemble dependence of the transient fluctuation theorem. *J. Chem. Phys.*, **113**, 3503–3509.

Searles, D.J. and Evans, D.J. (2004) Fluctuations relations for nonequilibrium systems. *Aust. J. Chem.*, **57**, 1119–1123.

Trepagnier, E.H., Jarzynski, C., Ritort, F., Crooks, G.E., Bustamante, C.J., and Liphardt, J. (2004) Experimental test of hatano and Sasa's nonequilibrium steady-state equality. *Proc. Natl. Acad. Sci. U.S.A.*, **101**, 15038–15041.

Wang, G.M., Sevick, E.M., Mittag, E., Searles, D.J., and Evans, D.J. (2002) Experimental demonstration of violations of the second Law of thermodynamics for small systems and short time scales. *Phys. Rev. Lett.*, **89**, 050601.

4
The Dissipation Theorem

> … the divergences [in the virial expansion of transport coefficients] somehow implied that the whole classical picture of nonequilibrium statistical mechanics was wrong, that there was an essential nonanalytic, non-controlled feature in the theory, that defied 'Boltzmann's dream.'
>
> *(Cohen, 1990)*

4.1
Derivation of the Dissipation Theorem

We now derive the *dissipation theorem,* which shows that, as well as being the subject of the Evans–Searles transient fluctuation theorem (ESFT), the dissipation function is the central argument of both linear response theory (i.e., Green–Kubo theory) and nonlinear response theory. This theorem was first derived in 2008 (Evans, Searles, and Williams, 2008a,b); see also Williams and Evans (2008).

Taking the solution of the Lagrangian form of the phase continuity Eqs. (2.26) and (2.29), we can substitute for $f(\mathbf{\Gamma};0)$ using the definition of the time-integrated dissipation function (3.2), obtaining

$$\begin{aligned} f(S^t\mathbf{\Gamma};t) &= \exp\left[-\int_0^t ds\,\Lambda\,(S^s\mathbf{\Gamma})\right]f(\mathbf{\Gamma};0) \\ &= \exp\left[-\int_0^t ds\,\Lambda\,(S^s\mathbf{\Gamma})\right]f(S^t\mathbf{\Gamma};0)\exp\left[\int_0^t ds\,\Omega(S^s\mathbf{\Gamma})+\int_0^t \Lambda\,(S^s\mathbf{\Gamma})\,ds\right] \\ &= f(S^t\mathbf{\Gamma};0)\exp\left[\int_0^t ds\,\Omega(S^s\mathbf{\Gamma})\right] \end{aligned} \tag{4.1}$$

The first line is obtained from Eqs. (2.26) and (2.29). The second line substitutes for $f(\mathbf{\Gamma};0)$ using the definition of the dissipation function (3.2).

Fundamentals of Classical Statistical Thermodynamics: Dissipation, Relaxation and Fluctuation Theorems, First Edition. Denis J. Evans, Debra J. Searles, and Stephen R. Williams.

Equation (4.1) is valid for all $\mathbf{\Gamma}$, so we choose $\mathbf{\Gamma} \to S^{-t}\mathbf{\Gamma}$. Then, after this remapping, we get

$$\begin{aligned} f(\mathbf{\Gamma};t) &= f(\mathbf{\Gamma};0)\exp\left[\int_0^t ds\,\Omega(S^s S^{-t}\mathbf{\Gamma})\right] \\ &= f(\mathbf{\Gamma};0)\exp\left[-\int_0^{-t} ds'\,\Omega(S^{s'}\mathbf{\Gamma})\right] \end{aligned} \tag{4.2}$$

where the second equality is obtained by introducing $s' = s - t$. Replacing the dummy variables gives

$$\begin{aligned} f(\mathbf{\Gamma};t) &= f(\mathbf{\Gamma};0)\exp\left[-\int_0^{-t} ds\,\Omega(S^s\mathbf{\Gamma})\right] \\ &= f(\mathbf{\Gamma};0)\exp\left[\int_0^{t} ds\,\Omega(S^{-s}\mathbf{\Gamma})\right] \end{aligned} \tag{4.3}$$

This result shows that the forward time propagator for the N-particle distribution function $\exp[-iL(\mathbf{\Gamma})t]$ – see Eq. (2.32): $f(\mathbf{\Gamma};t) = \exp[-iLt]f(\mathbf{\Gamma};0)$ – has a very simple relation (backward in time) to the exponential time integral of the dissipation function. In fact, Eq. (4.3) is a simpler equation. It involves only functions and their path integrals, whereas Eq. (2.32) involves functions and exponential integrals of *operators*.

If we take Eq. (4.3) and differentiate it in time at a fixed point in phase space, we see that

$$\frac{\partial f(\mathbf{\Gamma};t)}{\partial t} = \Omega(S^{-t}\mathbf{\Gamma})f(\mathbf{\Gamma};t) = -iL(\mathbf{\Gamma})f(\mathbf{\Gamma};t) \tag{4.4}$$

where we have used Eq. (2.31) to relate the time derivative of $f(\mathbf{\Gamma};t)$ to the f-Liouvillean.

Equation (4.4) shows there is a very simple relation between the dissipation function and the f-Liouville operator.

In the case of adiabatic (i.e., unthermostatted) dynamics for an ensemble that is initially an equilibrium canonical ensemble, this result is equivalent to the distribution function derived by Yamada and Kawasaki (1967). However, Eq. (4.3) is much more general and, like the ESFT, can be applied to any initial ensemble and any time-reversible, and possibly thermostatted, dynamics that satisfies AI$\mathbf{\Gamma}$. For thermostatted dynamics driven by a dissipative field, Eq. (4.3) was first derived in 1985 (Morriss and Evans, 1985).

From Eq. (4.3) we can calculate nonequilibrium ensemble averages of physical phase functions in the Schrödinger representation:

$$\begin{aligned} \langle B(t)\rangle_{F_e, f(\mathbf{\Gamma};0)} &= \int_D d\mathbf{\Gamma}\, B(\mathbf{\Gamma})\exp\left[-\int_0^{-t} ds\,\Omega\,(S^s\mathbf{\Gamma})\right]f(\mathbf{\Gamma};0) \\ &= \left\langle B\,(0)\exp\left[-\int_0^{-t} ds\ \ \Omega\,(S^s\mathbf{\Gamma})\right]\right\rangle_{F_e, f(\mathbf{\Gamma};0)} \end{aligned} \tag{4.5}$$

Differentiating Eq. (4.5) with respect to time, we find that

$$\begin{aligned}\frac{d\langle B(t)\rangle_{F_e,f(\mathbf{\Gamma};0)}}{dt} &= \int_D d\mathbf{\Gamma}\, B(\mathbf{\Gamma})\Omega(S^{-t}\mathbf{\Gamma})f(\mathbf{\Gamma};t)\\ &= \int_D d\mathbf{\Gamma}\, B(S^t\mathbf{\Gamma})\Omega(\mathbf{\Gamma})f(\mathbf{\Gamma};0)\\ &= \langle B(t)\Omega(0)\rangle_{F_e,f(\mathbf{\Gamma};0)}\end{aligned} \tag{4.6}$$

If we integrate Eq. (4.6) in time, we can write the averages of physical phase functions (Section 2.5) in the Heisenberg representation as

$$\langle B(t)\rangle_{F_e,f(\mathbf{\Gamma};0)} = \langle B(0)\rangle_{f(\mathbf{\Gamma};0)} + \int_0^t ds\langle \Omega(0)B(s)\rangle_{F_e,f(\mathbf{\Gamma};0)} \tag{4.7}$$

On both sides of Eq. (4.7), the time evolution is governed by the full field-dependent, thermostatted equations of motion (Eq. (2.14)).

Definition

The derivation of Eq. (4.7) from the definition of the dissipation function (3.2) is called the *dissipation theorem* (Evans, Searles, and Williams, 2008a,b). This theorem is extremely general and allows the determination of the ensemble average of an arbitrary physical phase function under very general conditions. We require time-reversible autonomous dynamics, an initial distribution that is invariant under the time reversal map M^T, and ergodic consistency so that the dissipation function is nonsingular.

Like the ESFT, Eqs. (4.5) and (4.7) are valid arbitrarily far from equilibrium. Equation (4.5) can be obtained for time-dependent fields by including the explicit time dependence, but Eq. (4.7) cannot (Williams and Evans, 2008). As in the derivation of the ESFT, the only unphysical terms in the derivation are the thermostatting terms within the wall region. However, because these thermostatting particles can be moved arbitrarily far from the system of interest, the precise mathematical details of the thermostat are unimportant. If the number of degrees of freedom in the reservoir is much larger than that of the system of interest, the reservoir can be assumed to be in thermodynamic equilibrium. In this case, there is therefore no difficulty in defining the thermodynamic temperature of the walls.

If the reservoir or thermostat is comparable in size to the system of interest, then both the reservoir and the system of interest may be far from equilibrium. The temperature, which is implicit in the actual expression for the dissipation function, is the equilibrium thermodynamic temperature both systems will relax to if the dissipative field is set to zero and both systems are allowed to relax to thermal equilibrium – see Chapter 5. For isokinetic systems, this is equal to the kinetic temperature of the thermostat. For Nosé–Hoover thermostatted systems, this is the target temperature of the Nosé–Hoover thermostat – regardless of the value of the Nosé–Hoover time constant. The instantaneous kinetic temperature of the Nosé–Hoover thermostatted reservoir particles is dependent on the particular value of the Nosé–Hoover time constant. The *only* temperature these differently thermostatted systems have in common is the equilibrium thermodynamic temperature they would each relax to, if they were allowed to do so.

4.2 Equilibrium Distributions are Preserved by Their Associated Dynamics

Equation (4.4) shows that if at time t, the dissipation is nonzero anywhere in the phase space domain, the distribution function $f(\mathbf{\Gamma};t)$ is time dependent:

$$\exists \mathbf{\Gamma} \in D st\, \Omega(S^t\mathbf{\Gamma}) \neq 0 \Rightarrow \frac{\partial f(\mathbf{\Gamma};t)}{\partial t} = \Omega(S^{-t}\mathbf{\Gamma})f(\mathbf{\Gamma};t) \neq 0 \tag{4.8}$$

and it cannot be an equilibrium distribution function. Conversely, if the distribution function is an equilibrium distribution at $t=0$ and the system evolves under zero-field dynamics in contact with a thermostat, then from Eqs. (3.30) and (4.3)

$$f_{\text{eq}}(\mathbf{\Gamma};t) = f_{\text{eq}}(\mathbf{\Gamma};0) = f_{\text{eq}}(\mathbf{\Gamma}), \quad \forall!\mathbf{\Gamma} \in D, \ \forall t \tag{4.9}$$

The distribution function will remain an equilibrium distribution function forever. Using Eq. (4.3)

$$\frac{\partial f_{\text{eq}}(\mathbf{\Gamma};t)}{\partial t} = \Omega_{\text{eq}}(S^{-t}\mathbf{\Gamma})f(\mathbf{\Gamma};t) = 0, \quad \forall!\mathbf{\Gamma} \in D, \ \forall t \tag{4.10}$$

Equation (4.10) in turn implies

$$\Omega_{\text{eq}}(\mathbf{\Gamma}) = 0, \quad \forall!\mathbf{\Gamma} \in D \tag{4.11}$$

where we have used Eq. (4.10) with $t=0$. So our definition of equilibrium involving path integrals of dissipation Eqs. (3.30) and (3.31) is equivalent to Eq. (4.11), which says that for equilibrium distributions the instantaneous dissipation must be zero everywhere. If you start with an equilibrium distribution, Eq. (4.10) implies that the distribution stays an equilibrium distribution for all time provided the dynamics remains that used in the definition of the dissipation function.

Furthermore, using Eq. (4.8) the only unchanging distribution functions are equilibrium distributions where the instantaneous dissipation is identically zero everywhere in the ostensible phase space domain. Thus distributions that, over some specified domain D, are at equilibrium with respect to their specified dynamics are time-independent at every point in phase space!

Definition

This gives us a new definition of equilibrium systems. *Equilibrium systems* are those combinations of dynamics and phase space distribution that satisfy Eq. (4.11). This is a simple restatement of our original definition (3.30), which involved time integrals of dissipation. Our new definition involves the instantaneous dissipation.

Notes:

- Although the partial derivative of the equilibrium distribution function with respect to time is zero, the streaming derivative of an equilibrium distribution function is *not* zero in general. As we will see in Chapter 6, for isochoric constant energy systems the streaming derivative is zero but for any equilibrium system that exchanges heat with its surroundings $df_{\mathrm{eq}}(t)/dt \neq 0$. For thermostatted equilibrium systems, the time-averaged streaming derivative is zero, however.
- Although the dissipation theorem shows that an equilibrium distribution $f_{\mathrm{eq}}(\boldsymbol{\Gamma})$ is preserved by its dynamics, we do not yet know whether an equilibrium distribution is unique or whether it is stable with respect to small perturbations. Neither do we know whether arbitrary initial distributions will relax towards equilibrium at long times. We will return to discuss these issues in Chapter 5.
- Equation (4.3) shows that for all *nonequilibrium* deterministic systems the N-particle distribution function has explicit time dependence: $f_{\mathrm{ne}}(\boldsymbol{\Gamma}; t)$. This automatically means that nonequilibrium steady-state distributions *cannot* be written in a closed, time-stationary form:

$$f_{\mathrm{ne}}(\boldsymbol{\Gamma}) \neq \frac{\exp[-F(\boldsymbol{\Gamma})]}{\int_D d\boldsymbol{\Gamma}\ \exp[-F(\boldsymbol{\Gamma})]} \tag{4.12}$$

  for some real $F(\boldsymbol{\Gamma})$. If Eq. (4.12) were possible, we would have

$$\frac{\partial f_{\mathrm{ne}}(\boldsymbol{\Gamma})}{\partial t} = 0 = \Omega(S^{-t}\boldsymbol{\Gamma}) f_{\mathrm{ne}}(\boldsymbol{\Gamma}, t), \quad \forall! t, \ \ \boldsymbol{\Gamma} \in D \tag{4.13}$$

  The only way this could happen would be if $\Omega(\boldsymbol{\Gamma}) = 0, \ \forall! \boldsymbol{\Gamma} \in D$. But this implies that the distribution is in fact an equilibrium distribution, which is a self-contradiction. Consequently Eq. (4.12) is correct.

  The Jaynes information theory approach (Jaynes, 1980) to nonequilibrium steady states hypothesizes closed forms like the right-hand side of Eq. (4.12) for nonequilibrium steady-state distributions. From Eq. (4.12), these can, *at most*, only be approximations! They cannot possibly be exact.

  In writing Eq. (4.11), we excluded the case where we discontinuously change the dynamics, thereby instantaneously changing the form of the equilibrium distribution. In such a case, the initial distribution is an equilibrium distribution for the prior dynamics ($t \leq 0$) but is a nonequilibrium distribution for the subsequent ($t > 0$) dynamics. The functional form of the dissipation function depends on the $t > 0$ dynamics.
- As noted in Section 2.5, in nonequilibrium steady states the distribution function collapses forever toward a steady-state attractor of lower dimension than that of the embedding phase space. Therefore, although averages of physical phase functions are time independent, in nonequilibrium steady states the distribution function and its associated Gibbs entropy Eq. (2.55) are not constant. The Gibbs entropy is, of course, not the average of a physical phase function; but rather it is an ensemble average of the logarithm of the

full N-particle phase distribution. Although the ensemble average of physical phase functions become time independent at sufficiently long times, the Gibbs entropy itself diverges linearly in time toward negative infinity.

4.3 Broad Characterization of Nonequilibrium Systems: Driven, Equilibrating, and T-Mixing Systems

Definition

A *driven* system is a system of interacting particles, possibly thermostatted in some way, subject to an external dissipative field F_e (or possibly asymmetric boundary condition). Consider a system that, for times up to zero, is in an equilibrium distribution with respect to the *zero-field* dynamics. The field-dependent dynamics satisfies AI$\mathbf{\Gamma}$. Because the zero-field system is at equilibrium with respect to the zero-field dynamics, the dissipative field is *solely* responsible for the dissipation for $t > 0$.

Definition

For driven systems the *dissipative factor*, $[\beta J](\mathbf{\Gamma})$, is defined by the equation,

$$\Omega(\mathbf{\Gamma}) \equiv -[\beta J](\mathbf{\Gamma})VF_e \tag{4.14}$$

where V is the system volume, and $[\beta J](\mathbf{\Gamma})$ is simply minus one times the instantaneous dissipation divided by the volume and the dissipative field. The factor of minus one is just conventional so that the dissipative flux J takes the same sign as the xy element of the pressure tensor when the dissipative field is the strain rate.

Definition

We often refer to Eq. (4.14) as the *primary dissipation function* for the external field F_e. When the field is zero, there is no dissipation.

Definition

The dissipative field could be a *mechanical field* appearing in the equations of motion (e.g., an electric field applied to an electrical conductor), or it could be the strain rate appearing in the SLLOD equations of motion (Section 2.3) when applied to a fluid.

Definition

The dissipative field could be a *thermodynamic field* (e.g., a velocity or temperature difference between moving walls that bound the system of interest). Thermodynamic fields are associated with boundary conditions. These boundary conditions do not usually appear in the actual equations of motion for the atoms or molecules comprising the system.

As we have seen, the SLLOD equations of motion when applied to a fluid system (Section 2.3) have the characteristics both of a mechanical dissipative process and a thermal transport process. The equations of motion refer explicitly to the field, but the boundary conditions also refer to the field. This points out that, although thermal and mechanical dissipative processes look profoundly different, at a deeper level there are similarities between the two types of field.

The SLLOD equations of motion, Section 2.3, are not autonomous but the nonautonomous terms rapidly decrease in magnitude with increasing system size in systems with short-range interatomic potentials such as the Lennard–Jones or WCA potentials (Evans and Morriss, 1990; Petravic, 2005; Bernardi, Brookes, and Searles, 2014).

Without loss of generality, we define the dissipative field so that the dissipation function is a *linear* functional of that field. If the dissipation is explicitly quadratic in some external physical variable, we just define the dissipative field to be that quadratic variable. The dissipative field in Eq. (4.14) is undefined up to some scalar factor. This has no serious mathematical consequences however, because this factor can be simply absorbed into the factor $[\beta J]$.

In order to specify $[\beta J](\boldsymbol{\Gamma})$ further, we need to look at the explicit form for the initial distribution and the dynamics. In Chapter 2, we showed that for Nosé–Hoover thermostatted driven systems whose equations of motion are given by Eq. (2.14), $\beta = 1/k_B T_{\text{th}}$. Here T_{th} was the target temperature of the Nose–Hoover thermostatted reservoir, which (as will be shown in Chapter 6), is equal to the equilibrium thermodynamic temperature that the entire system will relax toward if the dissipative field is set to zero and the entire system is allowed to relax toward equilibrium.

For ergostatted systems, β is not constant and is, instead, the reciprocal of the instantaneous kinetic temperature of the ergostatted particles times Boltzmann's constant. This kinetic temperature is *not* a constant of the motion for constant energy dynamics.

Analogous statements are made if the thermostat is, in fact, an isokinetic thermostat. In this case, β is the reciprocal of the constant kinetic temperature multiplied by Boltzmann's constant. Again, this kinetic temperature is the equilibrium thermodynamic temperature that the system will relax toward if the dissipative field is set to zero and the system is allowed to relax toward equilibrium.

For all driven systems that are at initially in equilibrium, Eq. (4.7) can be written as the transient time correlation function (TTCF) expression (Evans, Searles, and Williams, 2008a) for the thermostatted nonlinear response of the physical phase variable B to the dissipative field F_e:

$$\langle B(t)\rangle_{F_e,f(\boldsymbol{\Gamma};0)} = \langle B(0)\rangle_{f(\boldsymbol{\Gamma};0)} - V\int_0^t ds\langle[\beta J](0)B(s)\rangle_{F_e,f(\boldsymbol{\Gamma};0)} F_e \tag{4.15}$$

Definition

TTCF, Eq. (4.15), has been used frequently to compute the nonlinear transport behavior of systems over extremely wide ranges of the applied field (Morriss and Evans, 1987; Evans and Morriss, 1988; Borzsak, Cummings, and Evans, 2002; Todd and Daivis, 1999; Delhommelle, Cummings, and Petravic, 2005;

Delhommelle and Cummings, 2005; Pan and Mccabe, 2006; Brookes *et al.*, 2011). It is exact at arbitrarily far from equilibrium and for systems of arbitrary size. It applies to systems that are driven by mechanical fields that appear directly in the equations of motion and also to boundary-driven systems where it is the boundary conditions that prevent the system from being in equilibrium.

Definition

A system is said to be ΩT*-mixing*, over a phase space domain D, if ensemble averages over domain D of the TTCFs $\langle B(\mathbf{\Gamma}(s))\Omega(\mathbf{\Gamma}(0))\rangle$ appearing in Eq. (4.7) go to zero at long times sufficiently rapidly and their time integrals converge to a finite value as the integration time goes to infinity. In fact, Eq. (4.7) shows that ΩT-mixing is also a *necessary* condition for ensemble averages to be time-independent or stationary at long times.

Definition

Consider a system in which either $\langle A(0)\rangle$ or $\langle B(t)\rangle = 0,\ \forall t$ where $A(\mathbf{\Gamma}), B(\mathbf{\Gamma})$ are physical phase functions. Such a system is said to be *T-mixing* if TTCFs of $A(\mathbf{\Gamma}), B(S^t\mathbf{\Gamma})$ go to zero at long times, sufficiently rapidly for their infinite time integrals to converge to finite values: $\left|\int_0^\infty ds\ \langle A\ (0)\ B(s)\rangle\right| = \text{const} < \infty$. Obviously T-mixing systems are ΩT-mixing, and therefore T-mixing systems must relax to time-stationary states in the long time limit.

Definition

Consider a system in which either $\langle A(0)\rangle$ or $\langle B(t)\rangle = 0,\ \forall t$, where $A(\mathbf{\Gamma}), B(\mathbf{\Gamma})$ are physical phase functions. Such a system is said to be *weak T-mixing* if TTCFs of $A(\mathbf{\Gamma}), B(S^t\mathbf{\Gamma})$ go to zero at long times: $\lim_{t\to\infty}\langle A(0)B(t)\rangle = 0$.

If the decay of transient correlations takes place at a rate of $1/t$ or slower, weak T-mixing systems will not be T-mixing and will not be stationary at long times. All T-mixing conditions refer to physical phase functions.

Since the average of phase functions is not necessarily zero, that is, $\langle B(s)\rangle \neq 0$ for some B, the ΩT-mixing condition requires that

$$\langle \Omega(0)\rangle = 0 \tag{4.16}$$

and for driven systems

$$\langle [\beta J](0)\rangle = 0 \tag{4.17}$$

The dissipation function is odd under the time-reversal mapping and, since our initial distributions are always invariant under the time reversal mapping, Eq. (4.16) always holds. We can make some further remarks about Eq. (4.17). Because our systems are purely dissipative, Eq. (4.17) must be true. If it was not, then one could very slowly conduct a quasi-static change to the initial system that would change its thermodynamic state. In the absence of thermostats, if Eq. (4.17) were not true, we could slowly (quasi-statically) change the energy of the system. The Hamiltonian would be dependent on the external field. This violates our assumption that the dissipative flux is purely dissipative. This would violate the nonequilibrium partition identity. We will consider only fields that change the underlying thermodynamic state of the system in Chapter 8.

Definition

An *equilibrating* system is a system that evolves under zero-field dynamics, possibly in contact with some form of thermostat. Initially, the system is not in equilibrium with respect to the zero field dynamics. The initial form of the distribution is entirely responsible for dissipation.

If the dissipative field is nonzero, the ensemble-averaged and time-averaged dissipation is, as the second law inequality shows, always strictly positive, and the dissipation must be, to leading order, quadratic in the dissipative field. This means that the ensemble-averaged steady-state dissipation is analytic in the field, as expected for finite times in finite systems with continuous dynamics, and the system is driven as

$$\lim_{F_e\to 0}\langle[\beta J](t)\rangle_{F_e,f(\mathbf{\Gamma},0)} = O(F_e), \quad \forall t \tag{4.18}$$

It is possible that this leading order term vanishes because of some symmetry of the system, in which case the leading term would be cubic in the field.

For small fields and small systems, the averages of field-induced properties of the system are often swamped by noise from naturally occurring fluctuations. This makes direct calculation of the left-hand side of Eq. (4.15) problematic. This is particularly relevant in the calculation of the transport coefficient, which can be obtained from the ratio of the flux to the field. The TTCF can be applied at any field strength, and can even be zero, at which it reduces (essentially) to the Green–Kubo expression for the linear response (Kubo, 1966):

$$\lim_{F_e\to 0}\langle B(t)\rangle_{F_e,f(\mathbf{\Gamma};0)} = \langle B(0)\rangle_{f(\mathbf{\Gamma};0)} - V\int_0^t ds\langle[\beta J](0)B(s)\rangle_{F_e=0,f(\mathbf{\Gamma};0)} F_e \tag{4.19}$$

where the ensemble average on the right-hand side is an equilibrium ensemble average and the dynamics used to compute $B(s) \equiv B(S^s\mathbf{\Gamma})$ is the zero-field (possibly) thermostatted dynamics. This is in marked contrast to Eq. (4.15), where everything is computed with the dissipative field applied.

Definition

A system is said to be *mixing* if time correlation functions $\langle A(0)B(t)\rangle_\mu$ taken over a stationary distribution μ factorize in the long time limit: $\lim_{t\to\infty}\langle A(0)B(t)\rangle_\mu = \langle A\rangle_\mu\langle B\rangle_\mu$.

Weak T-mixing is a direct generalization of mixing for transient rather than stationary distributions. Mixing is for correlation functions in systems that have *stationary* averages of physical phase functions such as equilibrium or steady-state distributions.

Note: we could consider systems that are being driven by a dissipative field but are not initially at equilibrium with respect to the zero-field dynamics. For simplicity, we rarely consider such mixed systems in this book.

Definition

A stationary system is said to be *physically ergodic* if, in the long time limit, ensemble averages of physical properties are independent of the initial phase space distribution.

4.3.1 Two Corollaries of the Dissipation Theorem

Two consequences follow for systems that are T-mixing over the specified phase space domain. These systems have two properties:

1) They have *time-independent*, ensemble-averaged values for physical phase functions at long times.
2) They are *physically ergodic* over the specified phase space domain at long times.

These results are true for systems that are driven or equilibrating.

Property 1 is trivial. In fact, ΩT-mixing is a necessary and sufficient condition for stationarity at long times.

If the system is ΩT-mixing, obviously we have convergent integrals for Eq. (4.7) and a constant value for the long time ensemble-averaged value of all smooth physical phase variables. If we assume the system was not *physically ergodic*, then we could form time correlation functions involving the values of these physical phase functions with the time-zero dissipation. These correlation functions would never decay, violating the assumed T-mixing assumption. Thus T-mixing systems must be physically ergodic.

Why do we expect correlation functions (4.7) and (4.15) go to zero at long times? Two things happen. First, as we have seen, if the system is T-mixing, either Eq. (4.16) or (4.17) holds.

Second, in many (but not all!) systems, correlation functions of zero mean quantities go to zero at long times. (This is guaranteed if the system is weak T-mixing.) At late times, these systems lose "memory" of the initial value for the phase functions appearing in the correlation function. This loss of "memory" has *no* direct connection with Lyapunov instability or the Kolmogorov–Sinai entropy. In the linear response regime, the Fourier–Laplace transform of the decaying memory function in fact gives thermophysical information on how the system responds to periodic external fields at different frequencies (Evans and Morriss, 1990). In viscous systems, the decaying memory kernel characterizes the system's viscoelastic rheological properties.

From Eqs. (4.16) and (4.17), we have $\langle \Omega(0) \rangle = 0$ and many time-correlation functions decorrelate over time, and we have

$$\lim_{t\to\infty} \langle \Omega(0) B(t) \rangle = \langle \Omega(0) \rangle \lim_{t\to\infty} \langle B(t) \rangle = 0 \tag{4.20}$$

This will certainly occur if the system is weak T-mixing. The time correlation function appearing in Eq. (4.20) is not necessarily an equilibrium or steady-state correlation function. It may be a TTCF as in Eq. (4.7). Systems that do not lose correlations are generally integrable (e.g., undamped systems of harmonic oscillators).

Our definition of T-mixing is, in fact, stronger than Eq. (4.20). It requires that the correlations vanish sufficiently rapidly that the time integrals Eq. (4.7) converge to finite values. They must decay faster than t^{-1}. For equilibrium systems in two dimensions, autocorrelation functions of particle velocity, shear stress, and heat flux (all physical phase functions) evaluated in the limit of large system size are all thought to have divergent time integrals because of the so-called long time tails – see Section 8.7 of Hansen and Mcdonald (1986) for an elementary discussion. In three dimensions, the corresponding equilibrium autocorrelation functions are thought to decay asymptotically as $t^{-3/2}$, fulfilling the T-mixing convergence criterion.

Historically, there has been much interest in systems at the borderline of being mixing or T-mixing. The famous Fermi–Pasta–Ulam system (Gallavotti, 2008), which is a chain of anharmonic oscillators where the degree of anharmonicity can be controlled, are right at the border line of T-mixing. They are thought to be nonmixing.

References

Bernardi, S., Brookes, S.J., and Searles, D.J. (2014) System size effects on calculation of the viscosity of extended molecules. *Chem. Eng. Sci.*, **121**, 236–244.

Borzsak, I., Cummings, P.T., and Evans, D.J. (2002) Shear viscosity of a simple fluid over a wide range of strain rates. *Mol. Phys.*, **100**, 2735–2738.

Brookes, S.J., Reid, J.C., Evans, D.J., and Searles, D.J. (2011) The fluctuation theorem and dissipation theorem for poiseuille flow. *J. Phys. Conf. Ser.*, **297**, 012017.

Cohen, E.G.D. (1990) George E. Uhlenbeck and statistical mechanics. *Am. J. Phys*, **56**, 618–625.

Delhommelle, J. and Cummings, P.T. (2005) Simulation of friction in nanoconfined fluids for an arbitrarily low shear rate. *Phys. Rev. B*, **72**, 172201.

Delhommelle, J., Cummings, P.T., and Petravic, J. (2005) Conductivity of molten sodium chloride in an arbitrarily weak DC electric field. *J. Chem. Phys.*, **123**, 114505.

Evans, D.J. and Morriss, G.P. (1988) Transient-time-correlation functions and the rheology of fluids. *Phys. Rev. A*, **38**, 4142.

Evans, D.J. and Morriss, G.P. (1990) *Statistical Mechanics of Nonequilibrium Liquids*, Academic Press, London.

Evans, D.J., Searles, D.J., and Williams, S.R. (2008a) On the fluctuation theorem for the dissipation function and its connection with response theory. *J. Chem. Phys.*, **128**, 014504.

Evans, D.J., Searles, D.J., and Williams, S.R. (2008b) On the fluctuation theorem for the dissipation function and its connection with response theory [*J. Chem. Phys.* (2008), **128**, 014504], Erratum. *J. Chem. Phys.*, **128**, 249901.

Gallavotti, G. (2008) *The Fermi Pasta Ulam Problem*, Springer, Heidelberg.

Hansen, J.P. and Mcdonald, I.R. (1986) *Theory of Simple Liquids*, Academic Press.

Jaynes, E.T. (1980) The minimum entropy production principle. *Annu. Rev. Phys. Chem.*, **31**, 570–601.

Kubo, R. (1966) The fluctuation-dissipation theorem. *Rep. Prog. Phys.*, **29**, 255.

Morriss, G.P. and Evans, D.J. (1985) Isothermal response theory. *Mol. Phys.*, **54**, 629–636.

Morriss, G.P. and Evans, D.J. (1987) Application of transient correlation functions to shear flow Far from equilibrium. *Phys. Rev. A*, **35**, 792.

Pan, G. and Mccabe, C. (2006) Prediction of viscosity for molecular fluids at experimentally accessible shear rates using the transient time correlation function formalism. *J. Chem. Phys.*, **125**, 194527.

Petravic, J. (2005) Time dependence of phase variables in a steady shear flow algorithm. *Phys. Rev. E*, **71**, 011202.

Todd, B.D. and Daivis, P.J. (1999) A New algorithm for unrestricted duration nonequilibrium molecular dynamics simulations of planar elongational flow. *Comput. Phys. Commun.*, **117**, 191–199.

Williams, S.R. and Evans, D.J. (2008) Time-dependent response theory and nonequilibrium free-energy relations. *Phys. Rev. E*, **78**, 021119.

Yamada, T. and Kawasaki, K. (1967) Nonlinear effects in the shear viscosity of critical mixtures. *Prog. Theor. Phys.*, **38**, 1031–1051.

5
Equilibrium Relaxation Theorems

> One has therefore rigorously proved that, whatever the distribution of kinetic energy at the initial time might have been, it will, after a very long time, always necessarily approach that found by Maxwell.
>
> *(Boltzmann, 1872)*

5.1
Introduction

Understanding the relaxation of systems to equilibrium has been fraught with difficulties (Evans, Searles, and Williams, 2009a). The first reasonably general approach to this problem is the Boltzmann H-theorem. Beginning with the definition of the H-function, Boltzmann proved that the Boltzmann equation for the time evolution of the single-particle probability density in a uniform ideal gas implies a monotonic decrease in the H-function (Dorfman, 1999; Huang, 1963) – see the review by Lebowitz (1993) for a modern discussion of Boltzmann's ideas. However, there are at least two problems with Boltzmann's treatment. First, the Boltzmann equation is valid only for an ideal gas – its extension to higher densities has proven to be impossible (Cohen, 1990). Second, and more problematic, unlike Newton's equations, the Boltzmann equation itself is not time-reversal-symmetric. It is, therefore, completely unsurprising that it can be used to derive time-asymmetric results.

The middle of the twentieth century saw significant progress in ergodic theory with the proof (Sinai, 1976) that, since an autonomous Hamiltonian dynamical system preserves the microcanonical distribution, if the dynamics carried out within this microcanonical distribution is mixing, then in the long time limit, averages of physical phase functions should approach those of the uniform microcanonical distribution. In this chapter we will give this standard ergodic theory proof and a generalization that applies to thermostatted and/or barostatted relaxation. Curiously, these two proofs of relaxation do not refer explicitly to the dissipation function. Dissipation is, however, used to show that the equilibrium state exists and is stationary in time. Also, neither proof provides any details of the relaxation process. All they reveal is for finite systems that support mixing equilibrium states, long-time averages of physical phase functions, approach the equilibrium averages of those mixing equilibrium distributions.

Fundamentals of Classical Statistical Thermodynamics: Dissipation, Relaxation and Fluctuation Theorems, First Edition. Denis J. Evans, Debra J. Searles, and Stephen R. Williams.

Later we will use the dissipation theorem and a corollary of the Evans–Searles fluctuation theorem (ESFT), namely the second law inequality, to prove the relaxation to equilibrium of both autonomous Hamiltonian systems (Evans, Searles, and Williams, 2009b) and also of such systems in contact with a heat bath (Evans, Searles, and Williams, 2009a). We use these proofs to follow the details of the relaxation process. Our results extend the findings of modern ergodic theory, and they show the importance of dissipation in the process of relaxation toward equilibrium.

As an aside to the main logical development of this book, we prove that the negative logarithm of the canonical partition function is equal to the thermodynamic Helmholtz free energy divided by the thermodynamic temperature and Boltzmann's constant.

The results given in this chapter finally resolve the puzzle felt so keenly by R.C. Tolman 1938 (Tolman, 1979) concerning Boltzmann's postulate of equal *a priori* probabilities for the equilibrium state of autonomous Hamiltonian systems: "Although we shall endeavor to show the reasonable character of this hypothesis, it must nevertheless be regarded as a postulate which can be ultimately justified only by the correspondence between conclusions which it permits and the regularities in the behavior of actual systems which are empirically found." – Tolman (1979, p. 59).

5.2
Relaxation toward Mixing Equilibrium: The Umbrella Sampling Approach

It is known from ergodic theory that a finite, autonomous Hamiltonian system that preserves a mixing microcanonical equilibrium distribution will, from almost any initial state described by an phase space distribution $f(\mathbf{\Gamma};0)\delta(H(\mathbf{\Gamma})-E)$, eventually have averages of physical phase functions that relax toward their microcanonical equilibrium values (Sinai, 1976).

Definition

As mentioned in Section 4.3, a system is said to be *mixing* if for integrable, reasonably smooth physical phase functions, time correlation functions computed with respect to a stationary distribution factorize into products of averages computed with respect to the same distribution:

$$\lim_{t\to\infty}\left\langle A(\mathbf{\Gamma})B(S^t\mathbf{\Gamma})\right\rangle_\infty-\langle A(\mathbf{\Gamma})\rangle_\infty\langle B(\mathbf{\Gamma})\rangle_\infty=0 \tag{5.1}$$

Here, the brackets $\langle\cdots\rangle_\infty$ denote an ensemble average with respect to an invariant (i.e., time-stationary) probability distribution μ_∞. In the case that μ_∞ has a density $f(\mathbf{\Gamma};\infty)$, one may write

$$\langle A\rangle_\infty=\int d\mu_\infty(\mathbf{\Gamma})A(\mathbf{\Gamma})=\int d\mathbf{\Gamma}f(\mathbf{\Gamma};\infty)A(\mathbf{\Gamma}) \tag{5.2}$$

where $f(\mathbf{\Gamma};\infty)=|{}^{\partial\mu_\infty}/_{\partial\mathbf{\Gamma}}|$ and is a (dimensionless and normalized) distribution.

If $f(\mathbf{\Gamma};\infty)$ is not defined over the phase space of the system, one would write only the first equality $\langle A\rangle_\infty = \int d\mu_\infty(\mathbf{\Gamma})A(\mathbf{\Gamma})$, where $d\mu_\infty(\mathbf{\Gamma})$ is dimensionless and normalized.

Implicit in this definition is the fact that the invariant measure must be preserved by the dynamics. If it is not, $\lim_{t\to\infty}\left\langle B\left(S^t\mathbf{\Gamma}\right)\right\rangle_\infty \neq \langle B(\mathbf{\Gamma})\rangle_\infty$, because by definition $\mathbf{\Gamma}$ is sampled from $d\mu_\infty = d\mathbf{\Gamma} f(\mathbf{\Gamma};\infty)$ but $S^t\mathbf{\Gamma}$ will be sampled from some other distribution entirely. So mixing systems must, as a prerequisite, have an invariant measure that is preserved by the dynamics *and*, additionally, they must satisfy Eq. (5.1) with respect to this invariant distribution or measure.

We note that, if the system has nonzero angular momentum, no stationary long-time measure is possible (unless we transform to a non-inertial, co-rotating coordinate frame where Hamiltonian dynamics breaks down). So if angular momentum is conserved in our system, we must set it to zero.

The mixing property is a property of the stationary state of interest, in which observables take the average values denoted by $\langle\cdots\rangle_\infty$. It represents the fact that, in the macroscopically stationary state, correlations among time-evolving physical properties (measured by using averages of physical phase functions) decay in time. Therefore, in general, the mixing condition *would not appear* to guarantee relaxation to an invariant state. Mixing already *assumes* stationarity of the macrostate and its preservation by the system's dynamics regardless of whether it is reached asymptotically in time, as implied by our notation, or it is initially prepared in that state by some means.

Definition

Our version (Evans, Williams, and Rondoni, 2012) of the standard ergodic theory proof of relaxation for autonomous Hamiltonian systems begins by noting that the *microcanonical distribution*, $f_{\mu c}(\mathbf{\Gamma})$:

$$f_{\mu c}(\mathbf{\Gamma}) \equiv \lim_{dE\to 0}\frac{1}{\displaystyle\int_{E<H(\mathbf{\Gamma})<E+dE} d\mathbf{\Gamma}} \tag{5.3}$$

has zero dissipation for autonomous Hamiltonian dynamics and is therefore a time-stationary equilibrium distribution, preserved by the autonomous Hamiltonian dynamics – see Eq. (4.8). Here the domain, D, is the isoenergetic hypersurface with $E < H(\mathbf{\Gamma}) < E + dE$, and $dE \to 0$ and $f_{\mu c}(\mathbf{\Gamma})$ is nonzero on this domain. We assume that if our system is somehow inserted into this naturally invariant distribution, the ensemble of finite systems is *mixing*.

We will now give the standard proof that, if our ensemble is initially *not* distributed according to this distribution, the ensemble will, in fact, relax toward this distribution – at least for the purposes of computing time averages of low-order physical phase functions (Evans *et al.*, 2016). This last qualification is supremely important. In general the full N-particle relaxing phase space distribution is exceedingly highly structured. It collapses towards a fractal that at late times reflects unstable periodic orbits as the slowest decaying structures in the relaxing full phase space distribution. This would appear to make the theoretical analysis of phase space relaxation impossible. However because we are only interested in the relaxation of physical phase functions that are only functionals of extremely low order projected distributions (singlet pair and three-particle distributions – maybe four particle at most!) we can replace the full N-particle distribution with smooth relaxing distributions that have the same averages for physical properties. In the Green expansion (Green, 1952) of the full N-particle distribution the low order distributions are smooth and relax towards equilibrium. It is only the high order distribution functions that contain the fractal information. These high order distributions are unmeasureable in general and cannot be detected in the relaxing averages of physical phase functions. This allows us to replace the complex unrelaxing N-particle phase space distribution with the smooth counterpart assembled from the low order components of the Green expansion.

Note: If the dynamics has any constants of the motion, these should be fixed at values specified by appropriate delta functions as in Eq. (5.9).

We compute the time-dependent average of a physical phase function $A(\mathbf{\Gamma})$ for some smooth distribution function $f(\mathbf{\Gamma};t)$ which is defined in the domain D:

$$\begin{aligned}\langle A\rangle_t &= \int_D d\mathbf{\Gamma} A(\mathbf{\Gamma}) f(\mathbf{\Gamma};t)\\ &= \int_D d\mathbf{\Gamma} A(S^t\mathbf{\Gamma}) f(\mathbf{\Gamma};0)\end{aligned} \tag{5.4}$$

where the second equality is due to the equivalence of the Heisenberg and Schrödinger representations of phase space averages, and the notation $\langle A\rangle_t$ refers to an ensemble average with respect to the time-evolved distribution $f(\mathbf{\Gamma};t)$. In Eq. (5.4), stationarity is not assumed. However, since the dynamics is driven by an autonomous Hamiltonian, the energy is fixed.

Now we multiply and divide the last expression in Eq. (5.4) by the (necessarily finite!) ostensible volume of the phase space. This casts the first line in a form to which the mixing property can (perhaps) be applied:

$$\begin{aligned}\langle A\rangle_t &= \frac{1}{\int_D d\mathbf{\Gamma}} \cdot \int_D d\mathbf{\Gamma}\, A(S^t\mathbf{\Gamma}) f(\mathbf{\Gamma};0) \cdot \int_D d\mathbf{\Gamma}\\ &\equiv \left\langle A\left(S^t\mathbf{\Gamma}\right) f(\mathbf{\Gamma};0)\right\rangle_{\mu c} \cdot \int_D d\mathbf{\Gamma}\end{aligned} \tag{5.5}$$

We emphasize that, in order to derive Eq. (5.5), the ostensible phase space volume needs to be finite. This equation also requires that the asymptotic phase space density is defined at all points in D.

A few more words need to be said about $\left\langle A\left(S^t\mathbf{\Gamma}\right) f(\mathbf{\Gamma};0)\right\rangle_{\mu c}$. This function is an equilibrium microcanonical, cross-time correlation function. It results from the fact that for Hamiltonian dynamics, any time-dependent nonequilibrium ensemble average, say $\langle A\rangle_t$, equals a time-dependent nonequilibrium average $\left\langle A(S^t\mathbf{\Gamma}\right\rangle_0$ computed with respect to the initial distribution $f(\mathbf{\Gamma};0)$. It also assumes that $f(\mathbf{\Gamma};0) \equiv f_0(\mathbf{\Gamma})$ is a phase function (i.e. it is defined at all $\mathbf{\Gamma}$ and its time invariance is indicated by the '0'). In order for this to be possible we assume that the initial distribution is smooth; it could be an equilibrium distribution for a different dynamics. In such cases the distribution function behaves like a low order phase function, to which the mixing condition can be applied. We exclude the case where the initial distribution is a nonequilibrium steady-state distribution.

Using Eq. (5.1) and knowing that the microcanonical distribution is preserved by the autonomous Hamiltonian dynamics, we now take the long time limit:

$$\begin{aligned}\lim_{t\to\infty} \langle A\rangle_t &= \langle A(\mathbf{\Gamma})\rangle_{\mu c}\langle f(\mathbf{\Gamma};0)\rangle_{\mu c}\cdot\int_D d\mathbf{\Gamma}\\ &= \langle A(\mathbf{\Gamma})\rangle_{\mu c}\frac{1}{\int_D d\mathbf{\Gamma}}\int_D d\mathbf{\Gamma} f(\mathbf{\Gamma};0)\cdot\int_D d\mathbf{\Gamma}\\ &= \langle A(\mathbf{\Gamma})\rangle_{\mu c}.1 = \langle A(\mathbf{\Gamma})\rangle_{\mu c}\end{aligned} \tag{5.6}$$

We have used the mixing assumption Eq. (5.1) to allow us to factorize the naturally invariant (microcanonical) time correlation function into a product of two invariant (microcanonical) averages. Lastly, we use the normalization of the initial distribution function. Because distribution functions are normalized, we see that for any distribution and time $\langle f(\mathbf{\Gamma};t)\rangle_{\mu c} = \int_D d\mathbf{\Gamma} f(\mathbf{\Gamma};t)/\int_D d\mathbf{\Gamma} = 1/\int_D d\mathbf{\Gamma} = \exp[-S_G(E,N,V)/k_B]$, where we use the definition Eq. (2.55) and $f_{\mu c}(\mathbf{\Gamma}) = 1/\int_D d\mathbf{\Gamma}$ (see also Eq. (5.54) below). The microcanonical average of a normalized distribution function tells us nothing about that distribution. The average only tells us the Gibbs entropy of the microcanonical distribution used to calculate the average.

Note: we do not need to *assume* the existence of a stationary state, since the microcanonical distribution is indeed preserved by Hamiltonian dynamics. This is because (as already noted previously) the dissipation function is identically zero for autonomous Hamiltonian dynamics with an initial ensemble being the uniform microcanonical distribution and therefore the distribution is stationary using Eq. (4.4).

So $\langle A\rangle_t$ tends toward a microcanonical average, whatever physical phase function $A(\mathbf{\Gamma})$ or the initial probability density $f(\mathbf{\Gamma};0)$ one considers – as long as it lies on an energy hypersurface and the initial distribution is reasonably smooth. By definition, this amounts to a proof of relaxation *toward* the microcanonical equilibrium state denoted by $\langle\cdot\rangle_{\mu c}$.

Unless one starts at $t = 0$ with the microcanonical distribution, this proof shows that *averages* of low-order thermodynamic quantities *approach* microcanonical averages in the long time limit. The actual N-particle distribution never *becomes* the microcanonical distribution. At any time, no matter how large, we can always apply a time-reversal map and return (eventually!) to the initial distribution. As time increases in the relaxation process, the long-time N-particle distribution function becomes ever more tightly folded upon itself, never *becoming* the smooth microcanonical equilibrium distribution.

If it were in fact to eventually become *precisely* the microcanonical distribution one, could never return to the initial distribution by applying a time-reversal map. This gives a proof that the relaxation process cannot be complete in finite time.

We now generalize this derivation so that it applies to any dynamics that preserves a mixing equilibrium distribution $f_{eq}(\mathbf{\Gamma})$ (Evans *et al.*, (2016)). The phase space vector could be augmented by additional variables such as the Nosé–Hoover thermostat multiplier or the system volume to cover a variety of different systems (e.g., Nosé–Hoover thermostatted dynamics or Nosé–Hoover isothermal isobaric dynamics). However, we do not show this explicitly in our notation.

We write the equilibrium distribution as

$$f_{\text{eq},h}(\mathbf{\Gamma}) = \frac{\exp[-h(\mathbf{\Gamma})]}{\int_D d\mathbf{\Gamma}\,\exp[-h(\mathbf{\Gamma})]} \tag{5.7}$$

where $h(\mathbf{\Gamma})$ is some real integrable function of the (possibly augmented) phase space vector $\mathbf{\Gamma}$ defined over some domain D. Again, this initial distribution will need to be reasonably smooth. We compute the average of some physical phase space function $A(\mathbf{\Gamma})$ at some time t with an initial distribution $f_0(\mathbf{\Gamma}) \neq f_{\text{eq},h}(\mathbf{\Gamma})$:

$$\begin{aligned}
\langle A(t)\rangle_0 &= \int_D d\mathbf{\Gamma}\, A(S^t\mathbf{\Gamma}) f_0(\mathbf{\Gamma}) \\
&= \frac{\int_D d\mathbf{\Gamma}\, A(S^t\mathbf{\Gamma}) f_0(\mathbf{\Gamma}) \exp[h(\mathbf{\Gamma})] \exp[-h(\mathbf{\Gamma})]}{\int_D d\mathbf{\Gamma}\,\exp[-h(\mathbf{\Gamma})]} \int_D d\mathbf{\Gamma}\,\exp[-h(\mathbf{\Gamma})] \\
&= \left\langle A\left(S^t\mathbf{\Gamma}\right) f_0(\mathbf{\Gamma}) \exp[h(\mathbf{\Gamma})]\right\rangle_{\text{eq},h} \int_D d\mathbf{\Gamma}\,\exp[-h(\mathbf{\Gamma})] \\
&\xrightarrow[t\to\infty]{} \langle A(\mathbf{\Gamma})\rangle_{\text{eq},h} \left\langle f_0(\mathbf{\Gamma}) \exp[h(\mathbf{\Gamma})]\right\rangle_{\text{eq},h} \int_D d\mathbf{\Gamma}\,\exp[-h(\mathbf{\Gamma})] \\
&= \langle A(\mathbf{\Gamma})\rangle_{\text{eq},h} \frac{\int_D d\mathbf{\Gamma}\; f_0(\mathbf{\Gamma}) \exp[h(\mathbf{\Gamma})] \exp[-h(\mathbf{\Gamma})]}{\int_D d\mathbf{\Gamma}\,\exp[-h(\mathbf{\Gamma})]} \int_D d\mathbf{\Gamma}\,\exp[-h(\mathbf{\Gamma})] \\
&= \langle A(\mathbf{\Gamma})\rangle_{\text{eq},h} \int_D d\mathbf{\Gamma} f_0(\mathbf{\Gamma}) \\
&= \langle A(\mathbf{\Gamma})\rangle_{\text{eq},h}
\end{aligned} \tag{5.8}$$

where $\langle \cdots \rangle_{\text{eq},h}$ denotes an average with respect to the mixing equilibrium distribution given by Eq. (5.7). We note that all equilibrium distributions are stationary since they have zero dissipation everywhere in the allowed phase space. We also note that the "partition function" $\int d\mathbf{\Gamma}\,\exp[-h(\mathbf{\Gamma})]$ must be finite and nonzero:

otherwise the derivation cannot be completed. We also require a form of ergodic consistency: for all $\mathbf{\Gamma}$ s.t. $f_0(\mathbf{\Gamma}) \neq 0$, we require $\exp[-h(\mathbf{\Gamma})], \exp[h(\mathbf{\Gamma})] \neq 0$.

This latter point means that the derivation cannot be extended to thermostatted dissipative systems because the relevant "partition functions" and the relevant functions $\exp[-h(\mathbf{\Gamma})], \exp[h(\mathbf{\Gamma})]$ would be singular.

This generalized derivation of relaxation for arbitrary mixing equilibrium distributions can be applied to isokinetically thermostatted systems or Nosé–Hoover thermostatted systems as well as isothermal/isobaric systems. In each case, the relevant zero-dissipation equilibrium distribution has been known since the mid-1980s (Evans and Morriss, 1983)! We now have a proof of *relaxation* toward these smooth equilibrium distributions.

As elegant as the umbrella sampling "proof" of relaxation to equilibrium is, it reveals almost nothing of the relaxation process. It reveals nothing, for example, about the timescales for relaxation. Worse still, this proof cannot be extended to the question of relaxation to nonequilibrium steady states because of the lack of ergodic consistency. With this in mind, we construct a new proof of relaxation toward equilibrium using the notion of T-mixing. As we will see in Chapter 6, this new approach can indeed be applied to relaxation to nonequilibrium steady states.

The umbrella sampling proof of relaxation does show that the transient states connecting initial reasonably smooth distributions to the limiting equilibrium distribution *must* be ΩT-mixing, since ΩT-mixing is a necessary condition for systems to have stationary, long-time averages of physical properties. The existence of mixing equilibrium states that are preserved by a specified dynamics implies that the transients must be ΩT-mixing provided the initial states are also reasonably smooth. This is a somewhat surprising result. T-mixing transients imply that the limiting equilibrium state is mixing, but it is somewhat surprising that the mere existence of a mixing finite equilibrium state implies that the transients are at least ΩT-mixing!

5.3 Relaxation of Autonomous Hamiltonian Systems under T-Mixing (Evans, Searles, and Williams, 2009b)

We have already met the definition of the microcanonical distribution in Eq. (5.3). We repeat this definition in more detail here. From the definition of the dissipation function, it is trivial to see that, if the states are distributed as

$$
\begin{aligned}
f_{\mu c}(\mathbf{\Gamma}) &\equiv \lim_{dE\to 0} \frac{\delta(\mathbf{P})\delta(\mathbf{L})}{\displaystyle\int_{E<H(\mathbf{\Gamma})<E+dE} d\mathbf{\Gamma}\,\delta(\mathbf{P})\delta(\mathbf{L})} \\
&= \frac{1}{\displaystyle\int d\mathbf{\Gamma}} \quad \text{if} \quad \mathbf{\Gamma} \in D, \quad = 0 \text{ if } \mathbf{\Gamma} \notin D
\end{aligned} \tag{5.9}
$$

then under autonomous Hamiltonian dynamics the dissipation function is identically zero, everywhere in ostensible phase space, D, which is a limitingly thin energy shell, E, and zero total linear, $\mathbf{P}$, and total angular momentum $\mathbf{L}$. Basically, the linear and angular momenta are constants of the motion and the phase space expansion factor is also zero.

The distribution function in Eq. (5.9) is therefore an equilibrium distribution function. It is referred to as the equilibrium *microcanonical distribution* $(f_{\mu c}(\boldsymbol{\Gamma}))$ – see Eq. (5.3). Within this ostensible domain D, T-mixing systems have no physical phase functions that are constants of the motion. Later we will prove this statement from the T-mixing definition.

Of course, if the particular Hamiltonian with which we are dealing contains more symmetries than those discussed here, there will be additional physical phase functions that are constants of the motion. These should be handled by inserting additional delta functions into the microcanonical distribution Eq. (5.9) so that the ostensible phase space is constrained to a fixed value for these additional constants of the motion.

Mixing is closely related to, but subtly different from, the T-mixing condition. The equilibrium relaxation theorems require sufficiently fast rates of correlation decay. This is made part of the definition of T-mixing itself.

Definitions

In Section 4.3, we introduced the definitions of ΩT*-mixing*, *T-mixing*, and *weak T-mixing*. Because these definitions are very important, we remind the reader of these definitions once again. ΩT*-mixing* assumes that for a real, low-order, reasonably smooth physical phase function $A(\boldsymbol{\Gamma})$

$$\int_0^\infty ds \langle \Omega(\boldsymbol{\Gamma}) A(S^s\boldsymbol{\Gamma})\rangle_0 = L_0 \in \Re \tag{5.10}$$

where L_0 is real and finite, $\Omega(\boldsymbol{\Gamma})$ is the instantaneous dissipation at the phase $\boldsymbol{\Gamma}$ and $A(S^s\boldsymbol{\Gamma})$ is the phase function $A(\boldsymbol{\Gamma})$ evaluated at the time-evolved phase $S^s\boldsymbol{\Gamma}$. In contradistinction to the well-known mixing condition of ergodic theory, the T-mixing condition considers time correlation functions referred to the *initial* state, here denoted by $\langle\cdot\rangle_0$, where the distribution of phases is usually known.

T-mixing systems have the property that

$$\int_0^\infty ds\ \langle \delta A(\boldsymbol{\Gamma})\, \delta B(S^s\boldsymbol{\Gamma})\rangle_0 = L_0 \in \Re \tag{5.11}$$

where $\delta B(S^t\boldsymbol{\Gamma}) \equiv B(S^t\boldsymbol{\Gamma}) - \left\langle B\left(S^t\boldsymbol{\Gamma}\right)\right\rangle_0$ is a zero-mean physical phase function.

As in Section 4.3, the *weak T-mixing condition*, which looks very similar to the mixing condition, is

$$\lim_{t\to\infty} \left[\left\langle A(\boldsymbol{\Gamma}) B(S^t\boldsymbol{\Gamma})\right\rangle_0 - \langle A(\boldsymbol{\Gamma})\rangle_0 \left\langle B\left(S^t\boldsymbol{\Gamma}\right)\right\rangle_0\right] = 0 \tag{5.12}$$

where $A(\boldsymbol{\Gamma}), B(\boldsymbol{\Gamma})$ are low-order physical phase functions. The main difference between weak T-mixing, Eq. (5.12), and standard mixing, Eq. (5.1), lies in the fact that the second factor in the second term inside square brackets in Eq. (5.12) is not time-*independent*. It takes the form $\langle B(S^t\mathbf{\Gamma})\rangle_0 = \langle B(\mathbf{\Gamma})\rangle_t$, and hence it cannot be taken out of the limit. This time dependence is a reflection of the fact that the ensemble averages in Eq. (5.12) are taken with respect to the initial distribution rather than an invariant (*presumed*) long-time distribution.

For each of the T-mixing conditions (5.10–5.12), the relevant probability distribution is not the invariant one; it is the initial ensemble $d\mu_0(\mathbf{\Gamma}) = d\mathbf{\Gamma} f(\mathbf{\Gamma}; 0)$, whose averages are denoted by $\langle\cdot\rangle_0$. Mixing Eq. (5.1) and weak T-mixing Eq. (5.12) do not say anything about the rate of convergence to a stationary state or even whether such convergence actually occurs.

We obviously exclude the constants of the motion inherent in the Hamiltonian symmetries from being possible phase functions in Eqs. (5.11) and (5.12) (i.e., $A(\mathbf{\Gamma}),\ B(\mathbf{\Gamma}) \neq H_0(\mathbf{\Gamma}),\ P_\alpha(\mathbf{\Gamma}),\ L_\alpha(\mathbf{\Gamma}),\ \alpha = x, y, z$) since each of these variables is obviously a constant of the motion. So our ostensible phase space domain D is some specified physical volume on an energy hypersurface[1] with zero linear and angular momentum. The zero linear momentum condition could be relaxed, but the total angular momentum must be fixed at zero.

If the space is orientationally isotropic, the total angular momentum is a constant of the motion and, for reasons that are rather obvious, the system cannot possibly be T-mixing Eq. (5.11). When viewed from an inertial coordinate frame, the measure required for mixing Eq. (5.1) cannot be time-invariant but rather will be periodic. Likewise, the integrals required for the T-mixing property Eq. (5.11) will not in general converge but may also be periodic functions of the integration time. Rotating systems may, however, be weak T-mixing Eq. (5.12).

In a T-mixing system, there can be no nontrivial physical constants of the motion other than those inherent in the Hamiltonian symmetries. If there were such constants, we could form transient time correlation functions that violated Eq. (5.11). The fixed values of the various constants of the motion must be chosen to provide an inertial coordinate frame within which we can construct a Hamiltonian dynamical system.

All weak T-mixing systems are physically ergodic over the ostensible phase space because, if the phase space broke up into nonintersecting phase space subdomains characterized by different macroscopic averages for low-order phase functions, we could form constants of the motion depending on whether a system was on one subdomain or another. These subdomain occupation numbers could then be substituted as $A(\mathbf{\Gamma})$ in Eq. (5.12), thereby violating the weak T-mixing condition.

If the relevant time correlation functions (5.10) and (5.3) decay asymptotically as t^{-1} or more slowly, the system may be weak T-mixing Eq. (5.12) but cannot be ΩT-mixing or T-mixing Eqs. (5.10) and (5.11). In contradistinction to mixing Eq. (5.1), if a system is T-mixing Eq. (5.11) (or even ΩT-mixing Eq. (5.10)), it *must* relax to a time-stationary state at long times, whether or not this state is characterized by a smooth probability density $f(\mathbf{\Gamma}; \infty)$. If a system is weak T-mixing, but

1) This hypersurface is defined as a limitingly thin energy shell as in Eq. (5.3) rather than a true hypersurface H=E (Thompson, 1972).

not ΩT-mixing, relaxation to an invariant state from a noninvariant initial state will not occur. Generation of this weak T-mixing invariant state must be made by insertion into that state – say by a Monte-Carlo stochastic process.

In general, it is exceedingly difficult to prove that a given system is mixing and perhaps even harder to prove whether it is T-mixing. However, because of the many properties of T-mixing systems, it is easy to perform numerical/empirical tests of whether a system is T-mixing.

We now give a proof of relaxation to the stationary state based on the T-mixing condition (5.11) (Evans, Searles, and Williams, 2009b; Reid *et al.*, 2013). From the T-mixing assumption, there can be no low-order constants of the motion other than the trivial ones, the internal energy H_0 and the linear and angular momenta $\mathbf{P}, \mathbf{L}$, which are assumed to take on fixed values of $E, \mathbf{0}, \mathbf{0}$, respectively.

If we consider *any* reasonably smooth, deviation from the microcanonical form Eq. (5.9) generated by a real-valued integrable physical deviation function, $g(\mathbf{\Gamma})$ that is even in the momenta and differentiable,

$$f_g(\mathbf{\Gamma}) = \begin{cases} \dfrac{\exp\left[-g\left(\mathbf{\Gamma}\right)\right]}{\displaystyle\int d\mathbf{\Gamma}\ \exp[-g(\mathbf{\Gamma})]}, \mathbf{\Gamma} \in D \\ 0, \mathbf{\Gamma} \notin D \end{cases} \tag{5.13}$$

the dissipation function will not vanish and we have

$$\Omega(\mathbf{\Gamma}) = \dot{g}(\mathbf{\Gamma}) \tag{5.14}$$

where $\dot{g}(\mathbf{\Gamma}) \equiv \dot{\mathbf{\Gamma}} \cdot \partial g(\mathbf{\Gamma})/\partial \mathbf{\Gamma}$ denotes the time derivative.

Since the system is T-mixing, if $g(\mathbf{\Gamma}) \neq 0$, then $\Omega(\mathbf{\Gamma}) \neq 0$ because $g(\mathbf{\Gamma})$ cannot be a constant of the motion and the strict form of the second law inequality applies.

Definition

In Eq. (5.13), the physical phase function $g(\mathbf{\Gamma})$, which is even in the momenta, is termed a *deviation function.*

The strict second law inequality states that the ensemble average of the time integral of the dissipation from 0 to some time t is positive for all values of t. It is only equal to zero if the system is at equilibrium and $g(\mathbf{\Gamma}) = 0,\ \forall \mathbf{\Gamma}$. Thus for finite values of the deviation function g, we have

$$\left\langle \Omega_t \right\rangle_0 = \left\langle g\left(S^t\mathbf{\Gamma}\right) - g(\mathbf{\Gamma}) \right\rangle_0 \equiv \langle \Delta g\,(t) \rangle_0 > 0, \quad g(\mathbf{\Gamma}) \neq 0,\ \ \forall t > 0 \tag{5.15}$$

Thus, if there is *any* deviation from the equilibrium distribution Eq. (5.9), the dissipation function will not vanish (because there are no other low-order constants of the motion) and, further, the ensemble average of the time integrated dissipation function must be *positive.* In fact

$$\langle \Delta g\,(t) \rangle_0 = \int_0^{\infty} dA\,A(1 - e^{-A})p(\Delta g(t) = A) > 0, \quad \forall t,\ g(\mathbf{\Gamma}) \neq 0 \tag{5.16}$$

If there is any nonzero dissipation, $\Delta g(t) \neq 0$, the ensemble-averaged change in dissipation $\langle \Delta g\,(t) \rangle_0$ must be greater than zero. This means that for T-mixing systems with Hamiltonian dynamics, the smooth equilibrium distribution function is *unique* among distribution functions of the same or lower order, as $g(0)$, and is given by Eq. (5.9).

One can prove that the system must relax toward equilibrium by using the T-mixing property Eq. (5.10) and the dissipation theorem (Chapter 4) for the deviation function itself:

$$\lim_{t\to\infty} \langle g\,(t) \rangle_0 - \langle g\,(0) \rangle_0 = \lim_{t\to\infty} \left\langle \Omega_t \right\rangle_0 = \int_0^\infty ds \, \langle \dot{g}\,(0) g(s) \rangle_0 = \text{const} > 0$$
$$\Rightarrow \lim_{t\to\infty} \langle \dot{g}\,(t) \rangle_0 = \lim_{t\to\infty} \langle \Omega\,(t) \rangle_0 = 0 \tag{5.17}$$

Thus for the T-mixing systems treated here, in the long-time limit the ensemble-averaged instantaneous dissipation $\langle \dot{g}\,(t) \rangle$ is zero. In fact the ensemble averaged value of all phase functions that are odd under the time reversal mapping is in fact zero. This also means that the asymptotic distribution towards which the system is relaxing must be even under the time reversal mapping. Since we have already seen that the zero dissipation equilibrium state (even under the time reversal mapping) is unique, among distribution functions of the same, or lower, order as $g(0)$, the system must be relaxing toward that unique smooth equilibrium state.

Of course, for a system that is not initially at equilibrium, for any time no matter how large, the fine-grained phase space distribution is never given by the equilibrium state Eq. (5.9), because if this did happen, the system could never return to the initial distribution after the application of a time-reversal map.

This implies that for T-mixing autonomous Hamiltonian systems the relaxation to the true smooth equilibrium distribution function *must* take an infinite amount of time. Relaxation to equilibrium *cannot* occur in a finite time! Although the distribution function does not relax, the averages of low order phase functions (including the deviation function) calculated using the evolved distribution function become indistinguishable from those calculated using the smooth, unique equilibrium distribution.

5.4
Thermal Relaxation to Equilibrium: The Canonical Ensemble (Evans, Searles, and Williams, 2009a)

Consider a classical system of N interacting particles in a volume V. The microscopic state of the system is represented by a phase space vector of the coordinates and momenta of all the particles, $\{\mathbf{q}_1, \mathbf{q}_2, \ldots, \mathbf{q}_N, \mathbf{p}_1, \ldots, \mathbf{p}_N\} \equiv (\mathbf{q}, \mathbf{p}) \equiv \mathbf{\Gamma}$, where $\mathbf{q}_i, \mathbf{p}_i$ are the position and momentum of particle i. Initially (at $t = 0$), the microstates of the system are distributed according to a normalized probability distribution function $f(\mathbf{\Gamma}; 0)$. To apply our results to realistic systems, we separate the N-particle system into a system of interest and a wall region containing N_W

particles. Within the wall, a subset of N_{th} particles is subject to a fictitious thermostat. The thermostat employs a switch S_i, which controls how many and which particles are thermostatted, $S_i = 0;\ \ 1 \le i \le (N - N_{\text{th}}),\ S_i = 1;\ \ (N - N_{\text{th}} + 1) \le i \le N, N_{\text{th}} \le N_W$. We define the thermostat kinetic energy as

$$K_{\text{th}} \equiv \sum_{i=1}^{N} S_i \frac{p^2{}_i}{2m_i} \tag{5.18}$$

and write the equations of motion for the composite N-particle system as

$$\begin{aligned} \dot{\mathbf{q}}_i &= \frac{\mathbf{p}_i}{m_i} \\ \dot{\mathbf{p}}_i &= \mathbf{F}_i(\mathbf{q}) - S_i(\alpha \mathbf{p}_i + \mathbf{F}_{\text{th}}) \\ \dot{\alpha} &= \left[\frac{2K_{\text{th}}}{3\left(N_{\text{th}} - 1\right) k_{\text{B}} T_{\text{th}}} - 1 \right] \frac{1}{\tau^2} \end{aligned} \tag{5.19}$$

where $\mathbf{F}_i(\mathbf{q}) = -\partial\Phi(\mathbf{q})/\partial\mathbf{q}_i$ is the interatomic force on particle i, $\Phi(\mathbf{q})$ is the total interparticle potential energy, $-S_i\alpha\mathbf{p}_i$ is a deterministic, time-reversible Nosé–Hoover thermostat (Evans and Morriss, 1990) used to add or remove heat from the particles in the reservoir region through introduction of an extra degree of freedom described by α, T_{th} is the *target* parameter that controls the time-averaged kinetic energy of the thermostatted particles, and τ is the time constant for the Nosé–Hoover thermostat. The force $\mathbf{F}_{\text{th}} = (1/N_{\text{th}}) \sum_{i=1}^{N} S_i \mathbf{F}_i$ ensures that the macroscopic momentum of the thermostatted particles is a constant of the motion, which we set to zero.

Note that the choice of thermostat is reasonably arbitrary; for example, we could use some other choice of time-reversible deterministic thermostat, such as one obtained by use of Gauss' principle of least constraint (Evans *et al.*, 1983; Evans and Morriss, 1990), to fix K_{th} and arrive at essentially the same results. In order to simplify the notation, we introduce an extended phase space vector $\mathbf{\Gamma}^* \equiv (\mathbf{\Gamma}, \alpha)$ and from here on represent this implicitly using $\mathbf{\Gamma}$. In the absence of the thermostatting terms, the (Newtonian) equations of motion preserve the phase space volume so the system satisfies AI$\mathbf{\Gamma}$. The equations of motion for the particles in the system of interest are quite natural. The equations of motion for the thermostatted particles are supplemented with unnatural thermostat and force terms. Equations (5.18) and (5.19) are time-reversible, and heat can be either absorbed or given out by the thermostat.

For the Nosé–Hoover dynamics Eqs. (5.18) and (5.19), consider the initial distribution

$$f(\mathbf{\Gamma}; 0) \equiv f_c(\mathbf{\Gamma}) = \frac{\delta(\mathbf{P}_{\text{th}}) \exp[-\beta_{\text{th}} H_E(\mathbf{\Gamma})]}{\int_D d\mathbf{\Gamma}\, \delta(\mathbf{P}_{\text{th}}) \exp[-\beta_{\text{th}} H_E(\mathbf{\Gamma})]}, \quad \forall \mathbf{\Gamma} \in D \tag{5.20}$$

If $H_0(\mathbf{\Gamma})$ is the internal energy of the system, $H_E(\mathbf{\Gamma}) = H_0(\mathbf{\Gamma}) + {}^3\!/\!_2(N_{\text{th}} - 1) k_{\text{B}} T_{\text{th}} \alpha^2 \tau^2$ is the so-called extended Nosé–Hoover Hamiltonian, $k_{\text{B}} T_{\text{th}} \equiv \beta_{\text{th}}^{-1}$, and $\delta(\mathbf{P}_{\rm th}) \equiv \delta\left(\sum S_i p_{xi}\right)\delta\left(\sum S_i p_{yj}\right)\delta\left(\sum S_i p_{zk}\right)$ fixes the total momenta of the thermostatted particles in each Cartesian dimension, at zero.

Definition

We shall call the distribution in Eq. (5.20) the *canonical distribution* even though it includes extra degrees of freedom for the thermostat multiplier α.

Definition

In Eq. (5.20), $T_{\rm th} = \beta^{-1}/k_{\rm B}$ is called the *equilibrium thermodynamic temperature* of the canonical distribution of states Eq. (5.20).

It is easy to show that for this distribution Eq. (5.20) and the dynamics Eqs. (5.18) and (5.19), the dissipation function $\Omega_c(\boldsymbol{\Gamma})$ is identically zero at all points sampled by the canonical distribution:

$$\Omega_c(\boldsymbol{\Gamma}) = 0, \quad \forall \boldsymbol{\Gamma} \in D \tag{5.21}$$

Proof: From Eq. (5.20) and the definition of the dissipation function, we see that (see Appendix 2.A for how to evaluate Λ exactly)

$$\Omega_{c,t}(\boldsymbol{\Gamma}(0)) = \beta_{\rm th}[H_E(\boldsymbol{\Gamma}(t)) - H_E(\boldsymbol{\Gamma}(0))] + 3(N_{\rm th} - 1)\int_0^t ds\, \alpha(s) \tag{5.22}$$

Now, from the definition of the extended Hamiltonian and the equations of motion, we see that, if we take the time derivative of Eq. (5.22), we obtain

$$\Omega_c = \beta_{\rm th}[-2K_{\rm th}\alpha + 3(N_{\rm th} - 1)k_{\rm B}T_{\rm th}\alpha\dot{\alpha}\tau^2] - (3N_{\rm th} - 1)\alpha \tag{5.23}$$

Now, using the equation of motion for the thermostat multiplier, we see that

$$\begin{aligned}\Omega_c &= \beta_{\rm th}\left[-2K_{\rm th}\alpha + 3\left(N_{\rm th} - 1\right)k_{\rm B}T_{\rm th}\alpha\frac{2K_{\rm th} - 3(N_{\rm th} - 1)k_{\rm B}T_{\rm th}}{3(N_{\rm th} - 1)k_{\rm B}T_{\rm th}}\right] + 3(N_{\rm th} - 1)\alpha \\ &= \beta_{\rm th}[-2K_{\rm th}\alpha + \alpha[2K_{\rm th} - 3(N_{\rm th} - 1)k_{\rm B}T_{\rm th}]] + 3(N_{\rm th} - 1)\alpha \\ &= 0\end{aligned} \tag{5.24}$$

where we have used the fact that $k_{\rm B}T_{\rm th} \equiv \beta_{\rm th}^{-1}$. We note that in the proof we are using exact calculations. Often, approximations that are valid only in the large N limit are used in statistical mechanics. This calculation is exact for any arbitrary N.

We know from Section 4.2 that this initial (equilibrium) distribution is preserved by the dynamics Eqs. (5.18) and (5.19):

$$f(\boldsymbol{\Gamma}; t) = f_c(\boldsymbol{\Gamma}), \quad \forall \boldsymbol{\Gamma} \in D, \ \forall t \tag{5.25}$$

Since we know that Eq. (5.20) is an equilibrium distribution for the dynamics we consider, and since we also know that T-mixing systems are physically ergodic, we know from Section 4.3 that Eq. (5.20) is the unique smooth equilibrium distribution for this system. However, because of the importance of this point, we will explore the matter in greater detail.

Consider an arbitrary physical deviation from the canonical distribution

$$f(\mathbf{\Gamma};0) \equiv \frac{\delta(\mathbf{P}_{\text{th}})\exp[-\beta_{\text{th}}H_E(\mathbf{\Gamma}) - \gamma g(\mathbf{\Gamma})]}{\int_D d\mathbf{\Gamma}\,\delta(\mathbf{P}_{\text{th}})\exp[-\beta_{\text{th}}H_E(\mathbf{\Gamma}) - \gamma g(\mathbf{\Gamma})] \int} \tag{5.26}$$

where $g(\mathbf{\Gamma})$ is an arbitrary physical, integrable, real deviation function (phase function) and, since $f(\mathbf{\Gamma};0)$ must be an even function of the momenta, $g(\mathbf{\Gamma})$ must also be even in the momenta. Without loss of generality, we assume $0 \le \gamma$. The factor γ is a scale parameter that we can use to control the magnitude of the deviation from equilibrium.

For such a system Eq. (5.26) evolving under our dynamics Eqs. (5.18) and (5.19), the instantaneous dissipation function is

$$\Omega(\mathbf{\Gamma}) = \gamma \frac{\partial g(\mathbf{\Gamma})}{\partial \mathbf{\Gamma}} \cdot \dot{\mathbf{\Gamma}}(\mathbf{\Gamma}) = \gamma\, dg(\mathbf{\Gamma})/dt \tag{5.27}$$

Since $g(\mathbf{\Gamma})$ is even in the momenta, we know that

$$\langle \Omega(0) \rangle_g = \langle \gamma \dot{g}(\mathbf{\Gamma}) \rangle_g = 0 \tag{5.28}$$

where the subscript g on the ensemble average denotes the fact that the average is carried out over the initial deviated distribution (Eq. (5.26)).

Now, Eq. (5.27) implies

$$f(\mathbf{\Gamma};t) = \exp[-\gamma \Delta g(\mathbf{\Gamma}, -t)]\, f(\mathbf{\Gamma};0) \tag{5.29}$$

where $\Delta g(\mathbf{\Gamma}, t) \equiv g(S^t\mathbf{\Gamma}) - g(\mathbf{\Gamma})$. Because the system is T-mixing, there can be no physical constants of the motion additional to those specified in Eq. (5.11). Thus, if $g(\mathbf{\Gamma}) \neq 0$, there must be dissipation, and the distribution function cannot be a time-independent equilibrium distribution. Thus the smooth equilibrium distribution given by Eq. (5.20) is unique.

The dissipation function satisfies the strict second law inequality

$$\begin{aligned}\gamma \langle \Delta g(\mathbf{\Gamma}, t) \rangle_g &= \int_0^\infty dA\, A(1 - e^{-A}) p[\gamma \Delta g(\mathbf{\Gamma}, t) = A] \\ &> 0\end{aligned} \tag{5.30}$$

If $p[\gamma \Delta g(\mathbf{\Gamma}, t) = A]$ is nonzero for *any* $A > 0$, then $p[\gamma \Delta g(\mathbf{\Gamma}, t) = \pm A] > 0$ and the integrand in Eq. (5.30), as well as the integral, will be strictly positive. Thus in a T-mixing system if the initial distribution differs in any way from the canonical distribution, there will be dissipation and the ensemble-average of the time integral of the dissipation is *positive*. This remarkable result is true for an arbitrary $\gamma, g(\mathbf{\Gamma})$.

Summarizing, since the system is T-mixing, there is a *unique* time-symmetric, smooth equilibrium state characterized by being dissipationless everywhere in the phase space domain D. For the system considered here, that distribution is the canonical distribution (5.20). Thus we have derived an expression for the unique equilibrium state corresponding to the thermostatted equations of motion and shown that it takes on the standard form for the canonical distribution, modulo the facts that in the thermostatting region the momentum is a constant of the motion that is set to zero and that there is an extended degree of freedom for the thermostat.

If we start the system at time zero from a nonequilibrium distribution (5.26), we can ask the question how does the ensemble average of the deviation function change with time. Substitution of Eq. (5.30) into Eq. (5.29) gives

$$\langle \Delta g(\boldsymbol{\Gamma}, t)\rangle_g = \gamma \int_0^t ds \, \langle \dot{g}(0) g(s)\rangle_g > 0, \quad \forall t > 0 \tag{5.31}$$

where the ensemble averages are taken with respect to the initial nonequilibrium distribution function (5.26).

Because the initial distribution is an even function of the momenta, we know from Eq. (5.28) that the transient time correlation function appearing in Eq. (5.31) can be regarded as involving the product of two zero-mean phase variables (see Eq. (4.18)), and the T-mixing condition can be directly applied to the correlation function. Applying the T-mixing condition shows that the time integral on the right-hand side of Eq. (5.31) converges as $t \to \infty$. This implies that

$$\lim_{t\to\infty} \left| \langle \Delta g(\boldsymbol{\Gamma}, t+\tau)\rangle_g - \langle \Delta g(\boldsymbol{\Gamma}, t)\rangle_g \right| = 0 \tag{5.32}$$

This means that, as t becomes ever larger, the dissipation over a fixed time interval τ becomes ever smaller, and in the long-time limit the ensemble-averaged instantaneous dissipation vanishes. In fact the ensemble average of every odd phase function vanishes in the long time limit, implying the asymptotic distribution is even under the time reversal mapping. This implies that the system is relaxing toward its unique smooth equilibrium state, (as characterized by distribution functions of the same, or lower, order as $g(0)$) and

$$\lim_{t\to\infty} \gamma \langle \dot{g}(t)\rangle_{f(\boldsymbol{\Gamma},0)} = \lim_{t\to\infty} \langle \Omega(t)\rangle_{f(\boldsymbol{\Gamma},0)} = 0 \tag{5.33}$$

Equation (5.33) follows by differentiating Eq. (5.32) with respect to τ and then letting t increase without bound.

We have therefore proved that, subject to the conditions stated above, arbitrary initial nonequilibrium systems eventually relax, perhaps not monotonically, toward equilibrium (Evans, Searles, and Williams, 2009a). The distribution function itself continues to evolve for all time, but the averages of low-order phase functions become equal to those calculated with the smooth equilibrium distribution function.

Definition

We need to make an extremely important observation concerning the process of relaxation. The equilibrium canonical distribution (Eq. (5.20)) toward which the system with an arbitrary initial distribution (Eq. (5.26)) relaxes has an equilibrium thermodynamic temperature T_{th} that is identical (for systems of arbitrary size) to the *target* kinetic temperature of the Nosé–Hoover thermostat T_{th}. It is the *temperature of the underlying equilibrium state* the initial system will relax toward if it is T-mixing and it is so allowed.

From Eqs. (5.19) and (4.5), we see that

$$\langle B(t)\rangle_g = \langle B(0)\rangle_g + \gamma \int_0^t ds \langle \dot{g}(0) B(s)\rangle_g \tag{5.34}$$

and substituting $\dot{g}(\mathbf{\Gamma}) = B(\mathbf{\Gamma})$ we see that

$$\lim_{t\to 0^+} \langle \dot{g}(t)\rangle_g = \gamma \left\langle \dot{g}^2(0)\right\rangle_g > 0 \tag{5.35}$$

This proves that *initially*, on average, the system always moves toward, rather than away from, equilibrium. At later times, the system may move, for a short time, away from equilibrium (e.g., as in the case of an underdamped oscillator) but such movement is never enough to make the time-integrated, ensemble-averaged dissipation negative (or even zero). The time-integrated average dissipation from the initial state to any intermediate state (including the final equilibrium state) is strictly positive. At any sufficiently later instant in the relaxation process, the *instantaneous* dissipation may be negative. This shows that, in general, the relaxation process may not be monotonic in time. Such nonmonotonic relaxation is extremely common in nature.

Equation (5.35) implies another important point. Occasionally, one sees in the literature the correct observation that dissipative systems have phase space trajectories that are more stable than their time-reversed anti-dissipative conjugates – see, for example, William Thomson quote, Chapter 6 (Thomson, 1874) or Hoover (1999, p. 247). This comment on the relative mechanical stability is easily seen to be correct because, if we consider a nonequilibrium steady state, the sum of all the Lyapunov exponents must be negative. This implies that, for systems satisfying the conjugate pairing rule (Eq. (2.49)), the largest positive exponent for a steady state is smaller in magnitude than the largest positive exponent for an antisteady state. This is obvious because the largest positive exponent for an antisteady state is −1 times the value of the most negative exponent of a steady state, and the sum of the extremal exponents for a steady state that satisfies the conjugate pairing rule must be negative. For systems that do not satisfy the conjugate pairing rule, the Kolmogorov–Sinai entropy for the antisteady state is greater than that for the steady state.

However, this difference in the relative stability of steady state and antisteady state trajectories is irrelevant to whether the second "Law" of thermodynamics is being satisfied. Equation (5.35) shows that, on average, systems respond *immediately* in a direction favored by the second "Law." They do not rely on the slow buildup of instabilities before they begin to satisfy the second "Law." Indeed, the initial gradient of the response, Eq. (5.35), is an *equilibrium property* that is completely unrelated to Lyapunov exponents.

Since we now know that under the conditions specified here the system will at long times relax toward its unique equilibrium state, we therefore know the following:

$$\lim_{t\to\infty} \langle \dot{g}(t)\rangle_g = \langle \dot{g}(0)\rangle_g + \gamma \int_0^\infty ds \langle \dot{g}(0)\dot{g}(s)\rangle_g$$
$$\Rightarrow \gamma \int_0^\infty ds \langle \dot{g}(0)\dot{g}(s)\rangle_g = 0 \tag{5.36}$$

where the first term on the right-hand side of the top line is zero by Eq. (5.20) and the subscript g signifies that the initial ensemble is given by Eq. (5.26). Equation (5.36) is true for any deviation function that is even in the momenta.

Definition

We call Eq. (5.36) the *heat death equation* (Evans, Williams, and Rondoni, 2012). It shows that, for systems arbitrarily far from equilibrium initially, the infinite time integral of the transient autocorrelation function of fluxes of nonconserved quantities vanishes.

If we take the weak deviation limit where $\gamma \to 0$, we see that equilibrium time autocorrelation functions of fluxes of nonconserved quantities also vanish:

$$\lim_{\gamma\to 0}\lim_{t\to\infty} \langle \dot{g}(t)\rangle_g = \langle \dot{g}(0)\rangle_{\mathrm{eq}} + \gamma \int_0^\infty ds \langle \dot{g}(0)\dot{g}(s)\rangle_{\mathrm{eq}}$$
$$\Rightarrow \gamma \int_0^\infty ds \langle \dot{g}(0)\dot{g}(s)\rangle_{\mathrm{eq}} = 0 \tag{5.37}$$

This equation was first written down in 1963 by Zwanzig (Zwanzig, 1963; Berne, Boon, and Rice, 1966) and has been called the *ZBBR equation* (Evans, 1981, 1983).

In summary, we have demonstrated that, for any T-mixing Hamiltonian system of fixed volume and fixed number of particles in contact with a heat reservoir whose initial (nonequilibrium) distribution is even under time reversal symmetry,

- there is a *unique* dissipationless state, and this state has the canonical distribution (although a Nosé–Hoover thermostat was used in this derivation, essentially the same result is obtained with other thermostatting mechanisms such as a Gaussian isokinetic thermostat.);
- in T-mixing systems with decaying temporal correlations, the system relaxes toward canonical equilibrium;
- this relaxation toward equilibrium is *not* necessarily monotonic (we note that the Boltzmann H-theorem applied to uniform dilute gases implies a monotonic relaxation to equilibrium, thus the relaxation theorem allows for much more complex behavior as seen experimentally.);
- the time-integrated, ensemble-averaged dissipation satisfies the strict inequality $\langle \Delta g(\mathbf{\Gamma}, t)\rangle_g > 0$;
- the *initial* ensemble average response is always toward, rather than away from, equilibrium;
- the relaxation process cannot take place in finite time.

We have also shown quite generally that, for T-mixing dynamical systems obeying time reversible dynamics, equilibrium states have properties that are time-reversal-symmetric (i.e., probabilities of observing any set of trajectories and its conjugate set of antitrajectories are equal) if and only if the dissipation function is zero everywhere in phase space. If there is dissipation anywhere in the phase space, the distribution function is not time-independent and the system cannot be in equilibrium.

5.5
Relaxation to Quasi-Equilibrium for Nonergodic Systems

If the system is not T-mixing over the full phase space domain D, the system may split into nonergodic subdomains D_i $i=1,2$, and so on, each characterized by different ensemble averages for physical properties. If these states are individually T-mixing, then the two relaxation theorems given above (for Hamiltonian systems and for such systems in contact with a heat reservoir) still apply individually to each subdomain. The systems will still relax to either microcanonical or canonical equilibrium within each subdomain. Examples of such systems are relatively common, for example, solid or glassy systems. Many solid systems are not really completely relaxed to true thermodynamic equilibrium; their macroscopic physical properties are history-dependent – for example, work-hardened metals or metals that are rapidly quenched.

None of these systems is T-mixing over the ostensible phase space domain. However, all are expected to be T-mixing over the history-dependent phase space subdomains within which these solid samples are kinetically trapped. The topology of these subdomains can be incredibly complex. However, whatever the topology, because the macroscopic properties are stationary in time, such solids must be ΩT-mixing over the subdomain, and a particular solid sample is trapped within. Depending on whether such a system is in contact with a thermal reservoir or not, at long times such systems will relax toward microcanonical or canonical equilibrium within their particular phase space subdomain.

Quasi-equilibrium is very common in solids because their physical properties (essentially infinite shear viscosities and very low diffusion coefficients) mean that the full exploration of phase space is kinetically restricted – see the following for a more extensive discussion of relaxation to quasi-equilibrium states (Williams and Evans, 2007, 2008, 2010; Williams, Searles, and Evans, 2008).

5.6
Aside: The Thermodynamic Connection

This section is not necessary for the logical exposition of this book. It is included for those who already know classical thermodynamics. Statistical mechanics has been traditionally taught *assuming* the "laws" of thermodynamics. As we will see later in this book, the zeroth and second laws will be proved from mechanics in Sections 7.2 and 8.5, respectively, with an introduction to this development in Section 5.7.

We give a proof here, that the microscopic expressions defined below, Eqs. (5.39) and (5.40) on average, are indeed equal to the thermodynamic entropy and temperature, respectively (Reid *et al.*, 2013). We take as our starting point known expressions for the Galilei invariant energy and pressure to equal, on average, their thermodynamic counterparts. Energy and pressure are, as the first "law" of thermodynamics makes clear, completely mechanical in nature.

To begin, we note that from classical thermodynamics we have two equations for the entropy (S) in terms of the energy (U), the volume (V), and the pressure (p):

$$\left.\frac{\partial S}{\partial U}\right|_V = \frac{1}{T}$$
$$\left.\frac{\partial S}{\partial V}\right|_U = \frac{p}{T} \tag{5.38}$$

Consider the function $\widetilde{S}$ defined (up to an additive constant) as

$$\widetilde{S} = k_B \ln \int_D \delta(\mathbf{P})d\mathbf{\Gamma} \equiv k_B \ln V_\Gamma \tag{5.39}$$

where the integration domain D, is the limitingly thin energy shell. We can identify the internal energy U with the value of the Hamiltonian in a co-moving coordinate frame $H_0(\mathbf{\Gamma})$, because internal energy is the Galilei-invariant mechanical energy.

Consider a phase vector displacement in phase space $\mathbf{\Gamma}' = \mathbf{\Gamma} + d\mathbf{\Gamma}$, where $d\mathbf{\Gamma} = dU(\nabla_p H)/(\nabla_p H \cdot \nabla_p H)$ and $\nabla_p \cdots \equiv \left(\partial/\partial p_{x1}, \partial/\partial p_{y1}, \ldots, \partial/\partial p_{zN}, 0, 0, \ldots, 0\right) \ldots$ that is normal to the kinetic energy hypersurface $\sum_i p_i^2/2m = K_0$, and that (to leading order in N) changes the energy of any phase point $\mathbf{\Gamma}$ by a constant infinitesimal amount dU. Since the Jacobian of the transformation $J(\mathbf{\Gamma}) = \left|\partial\mathbf{\Gamma}'/\partial\mathbf{\Gamma}\right| = 1 + dU\nabla_p^2 H/(\nabla_p H \cdot \nabla_p H)$, it can be seen from Eq. (5.39) that (Butler *et al.*, 1998):

$$\left.\frac{\partial \widetilde{S}}{\partial U}\right|_V \equiv \frac{1}{\widetilde{T}} = \frac{3Nk_B}{2\langle K_0\rangle_{\mu c}} \equiv \frac{1}{\langle T_K(\mathbf{\Gamma})\rangle_{\mu c}} + O(1/N) \tag{5.40}$$

where the ensemble average is microcanonical and taken with respect to Eq. (5.9), and $T_K(\mathbf{\Gamma})$ is the instantaneous kinetic temperature. (Note: there are obviously infinitely many other phase space projections that one could use to move between two infinitely close energy hypersurfaces. These lead to infinitely many different phase functions whose microcanonical and canonical averages equate, in the thermodynamic limit, to the equilibrium thermodynamic temperature – see, for example, Eq. (5.55).)

If we now use the SLLOD equations (note that the SLLOD equations of motion give an exact description of arbitrary homogeneous flows – see Daivis and Todd (2006)) to accomplish an infinitesimal volume change at constant energy using an ergostat to fix the energy, we see that from the ergostatted equations of motion

$$dH_0 = dU = 0 = -pdV - 2K_0\alpha dt \tag{5.41}$$

where, from the SLLOD equations (2.21), p is the microscopic expression for the pressure in a bulk system which is spatially uniform over the range of intermolecular forces:

$$3pV = \sum_{i\in V} p_i^2/m - \frac{1}{2}\sum_{i\in V,\forall j} \mathbf{r}_{ij}\cdot\mathbf{F}_{ij} \tag{5.42}$$

where $\mathbf{r}_{ij} \equiv \mathbf{r}_j - \mathbf{r}_i$, $\mathbf{F}_{ij}$ is the force on particle i due to particle j, and α is the ergostat multiplier. (We assume that the intermolecular forces are limited to pair interactions only.) As Irving and Kirkwood showed (Irving and Kirkwood, 1950), this microscopic expression for p is easily identified with the microscopic mechanical force "across" a surface and is therefore, on average, equal to the thermodynamic pressure.

We also know from the phase continuity equation $df/dt = 3N\alpha f$, that the change in phase space volume dV_Γ caused by this constant energy volume change is

$$dV_\Gamma = -3N\langle\alpha\rangle_{\mu c} V_\Gamma dt \tag{5.43}$$

From our proposed microscopic equation for the entropy, (5.39), we see that

$$\left.\frac{\partial \widetilde{S}}{\partial V}\right|_U = \left\langle \frac{3Nk_B p}{2K_0}\right\rangle_{\mu c} = \left\langle \frac{p}{T_K}\right\rangle_{\mu c} = \frac{\langle p\rangle_{\mu c}}{\widetilde{T}} + O(1/N) \tag{5.44}$$

Comparing Eqs. (5.40), (5.44), and (5.38) and noting that the classical entropy is only defined up to an arbitrary constant, we conclude that S and $\widetilde{S}$ satisfy the same partial differential equation:

$$\left.\frac{\partial X}{\partial V}\right|_U \Big/ \left.\frac{\partial X}{\partial U}\right|_V = p \tag{5.45}$$

This means that up to an arbitrary additive constant the entropy and $\widetilde{S}(U, V)$ are the same function of U, V:

$$S(V, U) = \widetilde{S}(V, U) + O(1/N) + \text{const} \tag{5.46}$$

Note that $T, \widetilde{T}$, which are yet unresolved, both individually cancel from the two versions of Eq. (5.45) (when $X = S, \widetilde{S}$). Substituting the thermodynamic entropy into Eq. (5.44) and comparing with the second equation in Eq. (5.38) then shows that

$$T(V, U) = \widetilde{T}(V, U) + O(1/N) \tag{5.47}$$

The $O(1/N)$ corrections disappear in the thermodynamic limit where classical thermodynamics is valid. Having identified the microscopic expressions for the entropy and the temperature in equilibrium microcanonical systems, we can now apply the usual textbook arguments to derive the expressions for the Helmholtz free energy in canonical systems.

We can also give a microscopic expression for the Helmholtz free energy of equilibrium canonical systems directly.

We *postulate* that the Helmholtz Free energy $A(T, N, V)$ is the same function of the thermodynamic temperature that Q is of the Nose–Hoover target temperature T_{th} in Eq. (5.19) (Evans, Searles, and Williams, 2009a):

$$\begin{aligned} A(T = T_{\mathrm{th}}, N, V) &= Q(T_{\mathrm{th}}, N, V) + (1/N) \\ &\equiv -k_{\mathrm{B}} T_{\mathrm{th}} \ln \left[\int d\mathbf{\Gamma}\, \delta\left(\mathbf{P}_{\mathrm{th}}\right) \exp[-\beta_{\mathrm{th}} H_E(\mathbf{\Gamma})] \right] \end{aligned} \tag{5.48}$$

That is, when $T_{\mathrm{th}} = T$, the Helmholtz free energy $A(T)$ at the thermodynamic temperature T is equal to the value of the statistical mechanical expression $Q(T_{\mathrm{th}})$ that is defined in Eq. (5.48). From classical equilibrium thermodynamics, we note that the Helmholtz free energy satisfies the differential equation

$$U = A - T\frac{\partial A}{\partial T} \tag{5.49}$$

where U is the internal energy. Whereas if we differentiate Q, which is defined in (5.48), with respect to T_{th}, we see that

$$\left\langle H_0 \right\rangle = Q - T_{\mathrm{th}} \frac{\partial Q}{\partial T_{\mathrm{th}}} \tag{5.50}$$

Since $U = \left\langle H_0 \right\rangle$ and noting that when $T = T_{\mathrm{th}} = 0$, then $A(0) = U(0) = Q(0)$, we observe, treating T, T_{th} as integration parameters x, that A and Q satisfy the same differential equation: $U(x) = Y(x) - x^{\partial Y(x)}/_{\partial x}$, with the same initial $x=0$ condition and therefore $A(T) = Q(T_{\mathrm{th}})$ and our hypothesis Eq. (5.48) is proved. The pressure can be verified using the SLLOD equations, but since pressure is a mechanical property, obtaining a microscopic expression for pressure presents no difficulties.

Lastly, we derive a microscopic expression for the entropy of an equilibrium canonical system. First, we note from classical equilibrium thermodynamics

$$S = \frac{U - A}{T} \tag{5.51}$$

Substituting the microscopic expressions for both the Helmholtz Free energy and the internal energy into Eq. (5.51) for a canonical system gives

$$\begin{aligned} S_c &= \frac{\int d\mathbf{\Gamma}\, e^{-\beta H_0} H_0}{T \int d\mathbf{\Gamma}\, e^{-\beta H_0}} + k_{\mathrm{B}} \ln \left[\int d\mathbf{\Gamma}\, e^{-\beta H_0} \right] \\ &= \frac{-k_{\mathrm{B}} \int d\mathbf{\Gamma} \left(e^{-\beta H_0} \left[-\beta H_0 - \ln \int d\mathbf{\Gamma}\, e^{-\beta H_0} \right] \right)}{\int d\mathbf{\Gamma}\, e^{-\beta H_0}} \\ &= -k_{\mathrm{B}} \int d\mathbf{\Gamma} f_c(\mathbf{\Gamma}) \ln[f_c(\mathbf{\Gamma})] \end{aligned} \tag{5.52}$$

where $f_c(\mathbf{\Gamma})$ is given by the equilibrium canonical distribution (Eq. (5.20)).

Definition

The *Gibbs entropy* (see Eq. (2.55)) of a phase space distribution $f(\mathbf{\Gamma})$ is defined as

$$S_G \equiv -k_B \int d\mathbf{\Gamma} f(\mathbf{\Gamma}) \ln[f(\mathbf{\Gamma})] \tag{5.53}$$

The Gibbs entropy of the equilibrium canonical distribution (Eq. (5.20)) is the thermodynamic entropy of the equilibrium system in contact with a heat bath at the specified thermodynamic temperature. We note that in Eq. (2.57), as Gibbs knew, for autonomous Hamiltonian systems the Gibbs entropy is a constant of the motion.

The Gibbs entropy of a microcanonical distribution of states is the thermodynamic entropy of the isolated autonomous Hamiltonian system with internal energy U. We can calculate the Gibbs entropy of a microcanonical distribution of states:

$$\begin{aligned} S_{G,\mu c} &= -k_B \int_D d\mathbf{\Gamma} \frac{1}{\int_D d\mathbf{\Gamma}} \left\{ \ln[1] - \ln\left[\int_D d\mathbf{\Gamma}\right] \right\} \\ &= +k_B \ln\left[\int_D d\mathbf{\Gamma}\right] \end{aligned} \tag{5.54}$$

where D denotes the limitingly thin energy shell and zero momentum domain in phase space given in Eq. (5.9). Equation (5.54) is, of course, consistent with Eq. (5.39).

Lastly, we should make a comment seldom made in textbooks. We have derived a number of "standard" microscopic expressions for thermodynamic quantities. However, each such expression is not unique. For example, there are infinitely many different expressions for the equilibrium pressure or temperature. At equilibrium, you can even calculate the equilibrium thermodynamic temperature using an expression that is purely configurational (Butler *et al.*, 1998). If $\nabla_{\mathbf{q}} \equiv (\partial/\partial \mathbf{q}_1, \ldots, \partial/\partial \mathbf{q}_N)$, then

$$\frac{1}{k_B T} = \left\langle \frac{\nabla_{\mathbf{q}}^2 \Phi(\mathbf{q})}{|\nabla_{\mathbf{q}} \Phi(\mathbf{q})|} \right\rangle + O\left(\frac{1}{N}\right) \tag{5.55}$$

where $\Phi(\mathbf{q})$ is the interparticle potential energy of the system and $\mathbf{q}$ denotes all the Cartesian coordinates of all the particles in the system. Typically, what happens is that, away from equilibrium, these different expressions for equilibrium thermodynamic quantities each take on very different values, again pointing out how special the equilibrium state is.

5.7
Introduction to Classical Thermodynamics

We now return to the logical exposition of this book and give a quick derivation of some of the "laws" of classical equilibrium thermodynamics. A more detailed discussion will be given in Sections 7.2 and 8.5. We also take the opportunity to explain some of the unusual, in some cases unique, terminology and mathematics adopted by thermodynamicists over the last 150 years.

Typically, in the thermodynamic limit but very close to equilibrium, nonequilibrium averages of physical phase variables have negligible fluctuations – the standard deviation of averages of intensive phase functions scale like $N^{-1/2}$. These fluctuations can be ignored in the thermodynamic limit. Close to equilibrium, an intensive quantity like the hydrostatic pressure behaves like

$$\lim_{\dot{\varepsilon}\to 0} p(T_{\text{kin}}, \rho, \dot{\varepsilon}) = p(T = T_{\text{kin}}, \rho, \dot{\varepsilon} = 0) + O(\dot{\varepsilon}^2) \tag{5.56}$$

where T_{kin} is the kinetic temperature of the nonequilibrium system, T is the equilibrium thermodynamic temperature appearing in the equilibrium canonical distribution function (5.20), and $\dot{\varepsilon}$ is the dilation rate. In the so-called quasi-static limit, where rates of doing nonequilibrium processes go to zero, we can calculate changes using the equilibrium values for thermodynamic variables. This is allowed because integration *times* for processes scale like the reciprocals of the rates (e.g., $\dot{\varepsilon}^{-1}$), while the *errors* incurred by replacing the actual values with their equilibrium counterparts scale like the square of the rates (e.g., $\dot{\varepsilon}^2$ in Eq. (5.56) see Section 8.3 for another example discussed in more detail). This means that the nonequilibrium contributions to the integrated changes vanish as $O(\dot{\varepsilon})$ at sufficiently slow rates. Quasi-static changes are not mysterious processes. They are simply processes that take place in the infinitely slow *limit*, as in Eq. (5.56). This is, of course, *not* how processes always take place in the natural world!

Consider the Helmholtz free energy of our quasi-static canonical system in the thermodynamic limit ($N \to \infty$): $A = -k_B \ln\left[\int_D d\mathbf{\Gamma} \exp[-\beta H_0(\mathbf{\Gamma})]\right]$, see Eq. (5.38). Let our system be subject to a quasi-static cyclic change in the temperature. From our definition of the Helmholtz free energy of our equilibrium canonical system, we see that for quasi-static changes

$$\dot{A} = k_B \left\langle \frac{\partial \beta H_0}{\partial t} \right\rangle_C \tag{5.57}$$

where $\langle \cdots \rangle_C$ denotes a canonical average.

Now we compute the cyclic integral of the quasi-static rate of change in the Helmholtz free energy due to temperature changes in the thermodynamic limit:

$$\begin{aligned} {}_{\text{qs}}\oint dA &= {}_{\text{qs}}\oint d(U/T) \\ &= {}_{\text{qs}}\oint U d(1/T) + {}_{\text{qs}}\oint dQ/T \\ &= 0 \end{aligned} \tag{5.58}$$

The first line is obtained from Eq. (5.57) and noting that in the thermodynamic limit $H_0 = U$. In going from the first to the second line in Eq. (5.58), we use the chain rule and the fact that for this transformation, by construction, the change in the internal energy is caused solely by the exchange of heat.

In Eq. (5.58), we follow the conventional notation used in classical thermodynamics, where dX denotes a quasi-static so-called *virtual change* in the variable X. We could replace each dX by $\lim_{dX/dt\to 0} dX/dt$, but conventionally this has not been done.

The fact that the cyclic integral of the change in the Helmholtz free energy of our quasi-static (equilibrium) is zero is due to the fact that, by definition, in a cyclic quasi-static process, the system returns to its initial equilibrium state after one cycle. Similarly, both the internal energy and the temperature return to their original values upon completion of a quasi-static cycle, so ${}_{\mathrm{qs}}\oint Ud(1/T) = 0$.

Substituting this latter expression into Eq. (5.58) shows that

$$ {}_{\mathrm{qs}}\oint dQ/T = 0 \tag{5.59}$$

If we split this cyclic integral into two part cycles: $1 \to 2,\ \ 2 \to 1$, the fact that the cyclic integral is always zero means that, given a pathway $1 \to 2$, the integral for the return path back to 1 is independent of whether the return pathway includes an intermediate state 3 or a different intermediate state, say 4. Thus ${}_{\mathrm{qs}}\int_2^1 dQ/T$ is independent of the pathway between 2 and 1.

Definition

A thermodynamic *state function* is solely a function of the thermodynamic state variables, which for our system comprise the number of particles N, the thermodynamic temperature T, and the volume of the system V or equivalently the density ρ. The numerical value of a state function is independent of the pathway or history describing how it arrived at its current state.

Definition

This enables us to define the change in a state function called the *calorimetric entropy* $S(T, V)$ as

$$ S_2 - S_1 \equiv {}_{\mathrm{qs}}\int_1^2 đQ/T \tag{5.60}$$

The *strike* through the differential for the heat denotes the fact that the heat is not a state function, and many textbooks say that $đQ$ is an "imperfect" differential. Textbooks also sometimes state that the reciprocal of the absolute temperature is the "integrating factor for the heat, turning $đQ/T$ into a perfect differential: dS" – meaning that its integral Eq. (5.60) is a state function, namely the entropy S.

We can also see that this integrating factor is unique. It cannot be replaced by any other monotonic function of the temperature, say $1/T^3$. This is because it comes ultimately from the algebraic form for the canonical equilibrium distribution function (5.20), then to the associated form for its corresponding Helmholtz free energy Eq. (5.38), and then into its derivative Eq. (5.57). Since for T-mixing systems in contact with a heat reservoir, the equilibrium distribution is, as we proved, unique, so too is the "integrating factor" for the heat. So the integrating factor for the heat is, in fact, the instantaneous temperature of the quasi-static canonical distribution of states through which the system evolves.

Now, for a system subject to a change of volume and a transfer of heat, the conservation of energy shows that for quasi-static processes in the thermodynamic limit

$$dU = đQ - pdV \tag{5.61}$$

This is usually referred to as the *first "Law" of thermodynamics*, but it is really nothing but the statement of conservation of total mechanical energy. (Note: we could include an infinite variety of other forms of work in Eq. (5.61).) We only include the work performed by moving a piston against a pressure head, for simplicity.

Since the heat transfer is not a state function, we can use Eq. (5.60) to replace it with variables that are state functions. This gives us the fundamental statement of the combined first and second "laws" of classical thermodynamics for quasi-static changes in the thermodynamic limit:

$$dU = TdS - pdV \tag{5.62}$$

Definition

Equation (5.62) is termed the *Gibbs equation*. It was first written down in this form by Clausius. Gibbs then generalized it so that it could be applied to mixtures. Over time, the special case given here has also been called the *Gibbs equation.*

We note that, if we take the microscopic definition of the Gibbs entropy of a canonical system given in Eq. (2.55) (repeated in Eq. (5.53)), we can see that

$$\begin{aligned}
TS_{G,c} &= -k_B T \int d\boldsymbol{\Gamma} f_c(\boldsymbol{\Gamma}) \ln[f_c(\boldsymbol{\Gamma})] \\
&= \frac{-k_B T \int d\boldsymbol{\Gamma} \left(e^{-\beta H_0} \left[-\beta H_0 - \ln \int d\boldsymbol{\Gamma}\, e^{-\beta H_0} \right] \right)}{\int d\boldsymbol{\Gamma}\, e^{-\beta H_0}} \\
&= \frac{\int d\boldsymbol{\Gamma}\, e^{-\beta H_0} H_0}{\int d\boldsymbol{\Gamma}\, e^{-\beta H_0}} + k_B T \ln \left[\int d\boldsymbol{\Gamma}\, e^{-\beta H_0} \right] \\
&= U - A
\end{aligned} \tag{5.63}$$

where we have used the definition (5.48) for the Helmholtz free energy, which we repeat here:

$$A \equiv -k_B T \ln \left[\int d\boldsymbol{\Gamma}\, e^{-\beta H_0} \right] \tag{5.64}$$

The first term in the second last line of Eq. (5.63) is obviously the canonical equilibrium average of the energy (i.e., the internal energy U). Equation (5.63) shows the fundamental relationship between the internal energy and the Helmholtz free energy.

Equation (5.60) enables the entropy of any equilibrium state to be determined by carrying out a quasi-static change from an initial equilibrium state where the entropy is known. However, it cannot be used to determine the entropy of a nonequilibrium state. By construction, the final state must be at equilibrium. We therefore define an irreversible calorimetric entropy S_{ir}, which is no longer a state function:

$$S_{\mathrm{ir},2} - S_1 \equiv \int_1^2 đQ/T \tag{5.65}$$

We can now show that the instantaneous value of the rate of change of the Gibbs entropy defined in Eq. (2.55) is equal to the rate of change of the calorimetric entropy defined in Eq. (5.65). This is true for both quasi-static and irreversible processes. Both entropies are, apart from a constant that is independent of the thermodynamic state of the system, one and the same quantity.

Consider the instantaneous rate of change of the Gibbs entropy for a possibly *irreversible* process:

$$\begin{aligned}
\dot{S}_G(t) &= -k_{\mathrm{B}} \int d\boldsymbol{\Gamma}\,[1 + \ln(f(\boldsymbol{\Gamma};t))]\frac{\partial f(\boldsymbol{\Gamma};t)}{\partial t} \\
&= -k_{\mathrm{B}} \int d\boldsymbol{\Gamma} f \dot{\boldsymbol{\Gamma}} \cdot \frac{\partial[1+\ln(f)]}{\partial \boldsymbol{\Gamma}} \\
&= -k_{\mathrm{B}} \int d\boldsymbol{\Gamma} \dot{\boldsymbol{\Gamma}} \cdot \frac{\partial f}{\partial \boldsymbol{\Gamma}} = k_{\mathrm{B}} \int d\boldsymbol{\Gamma} f \frac{\partial}{\partial \boldsymbol{\Gamma}} \cdot \dot{\boldsymbol{\Gamma}} = -3N_{\mathrm{th}} k_{\mathrm{B}} \langle \alpha(t) \rangle \\
&= \frac{-\dot{Q}_{\mathrm{th}}(t)}{T} = \frac{\dot{Q}_{\mathrm{soi}}(t)}{T} = \dot{S}_{\mathrm{ir}}(t), \quad \forall t
\end{aligned} \tag{5.66}$$

In the last line $\dot{Q}_{\mathrm{soi}}(t)$ denotes the heat gain or loss for the system of interest. In going from line 3 to line 4 of Eq. (5.66), we have assumed that N_{th} particles are subject to an isokinetic thermostat in three Cartesian dimensions. In this case, the temperature in line 4 would be the kinetic temperature of the N_{th} particles, which may be some, or all, of the particles in the system. This makes no difference to the final result because in either case this kinetic temperature is equal in value to the equilibrium thermodynamic temperature T the entire system will relax to if it is so allowed. If the process is not quasi-static, the change in the entropy computed using Eq. (5.65) is *not* path-independent. This means that, if the process is irreversible, Eq. (5.60) cannot be used since the left hand-side of the equation is undefined if the pathway between states 1 and 2 is undefined.

We note that for a quasi-static process transforming a system from equilibrium state 1 to equilibrium state 2, $\Delta S_G = \Delta S_{\mathrm{ir}} = \Delta S_{\mathrm{eq}}$, where ΔS_{eq} is defined through Eq. (5.60). However, if the transformation is carried out irreversibly, $\Delta S_G = \Delta S_{\mathrm{ir}} \neq \Delta S_{\mathrm{eq}}$. This reflects the fact that only the equilibrium entropy is a state function. It also reflects the fact that the Gibbs entropy incorporates the details of the underlying distribution that will never reach the equilibrium distribution of state 2 in a finite time, although the physical properties of the system will relax to the equilibrium values to within any desired accuracy.

Definition

Equation (5.66) gives a generalized definition for the *rate of change of entropy*. It is valid for reversible and irreversible processes and shows that up to a state-independent, additive constant, the Gibbs and irreversible calorimetric entropies are equal. Their rates of change are instantaneously equal. The key to understanding Eq. (5.66) is our new definition of the temperature of the underlying equilibrium state that any nonequilibrium system would relax to if it was so allowed.

One could make infinitely different variations on the thermostatting mechanism. We could just thermostat one particle, or we could thermostat just the x-kinetic temperature of some or all of the particles, but the final result remains unchanged. In this sense, the factor $3N_{\text{th}}$ in line 2 of Eq. (5.66) is completely nominal. The last line gives the only physically relevant result. As we will see in Chapter 8, we could even use a Nosé–Hoover thermostat without changing the result.

In the Nosé–Hoover thermostat, we could even make the target kinetic temperature a time dependent function – Section 8.5. In this case, Eq. (5.66) is still valid except that the temperature must be the instantaneous value of the target kinetic temperature which, of course, is also the instantaneous value of the temperature of the underlying equilibrium state. This temperature is not instantaneously equal to the ensemble-averaged kinetic temperature for this system. This illustrates the unique properties of the temperature of the underlying equilibrium state.

Equation (5.66) provides a very simple example of how singular the entropy is for nonequilibrium systems. Suppose we start with a microcanonical equilibrium distribution for some ergostatted dynamics in the thermodynamic limit. We apply a dissipative field to the system in which the AIΓ condition holds. This means that Eq. (5.66) gives the rate of change of the entropy of the system of interest, even in the presence of the dissipative field. After a Maxwell time, we assume the system to relax into a nonequilibrium steady state (in Chapter 6, we show the T-mixing is a necessary and sufficient condition for relaxation to a nonequilibrium steady state).

In the steady state, the ensemble-averaged value of the ergostat multiplier is a constant independent of time. The entropy of the system diverges toward negative infinity at a constant rate. After some time t, we set the dissipative field back to zero and let the T-mixing system relax at constant energy *toward* microcanonical equilibrium. In this relaxation process, the entropy of the system does not change because it is now an autonomous Hamiltonian system. After a Maxwell time (or multiple Maxwell times), the averages of all physical phase functions take on (to arbitrary accuracy) the same microcanonical values they had at time zero, and yet the entropy of the final state can be made arbitrarily *more negative* than the initial entropy by simply extending the duration t of the nonequilibrium steady state to arbitrarily large values! In Appendix 5.A, we illustrate these same issues for a system subject to a Nosé–Hoover thermostat.

Although we have given meaning to temperature in non-quasi-static or irreversible processes, the entropy calculated for such processes is, unsurprisingly, *not* a state function. Only in the quasi-static *limit* is the entropy a state function (5.60).

The Gibbs entropy, which is equal to the irreversible calorimetric entropy, is a function of the unphysical N-particle distribution function. An equilibrium distribution can, as we have just seen, never be reached in finite times. This is why by introducing the T-mixing condition we concentrate on the relaxation of averages of *physical* phase functions, ignoring the fractal intricacies of N-particle phase space distribution functions.

The derivation of classical thermodynamics will be explained in greater detail in Chapter 8, where we will give a detailed account of how the changes in the state of the system take place. In Chapter 8, we will also prove some thermodynamic *inequalities* such as the Clausius inequality for a thermal reservoir. Superficially, some of these inequalities will look similar to the inequalities found in thermodynamic textbooks, but the meaning of many of the quantities is different. For example, we have already given a definition for the temperature appearing in the Clausius inequality for irreversible processes. It is the equilibrium thermodynamic temperature the system will relax toward, if it is so allowed. In Chapter 8, we will use Nosé–Hoover thermostats rather than the isokinetic thermostats used here. This also shows that the basic equalities of classical thermodynamics are independent of how the thermostatting is performed.

5.A Appendix: Entropy Change for a Cyclic Temperature Variation (Williams, Searles, and Evans, 2014)

Here we consider a system subject to a cyclic change in temperature to demonstrate the behavior and relationship between the dissipation and the thermodynamic and Gibbs entropies. Consider a thermostatted system at equilibrium at T_1, which is monitored for a period τ_1, then is decreased to temperature T_2 over a period τ_2, maintained that temperature for a period τ_3, then warmed back to T_1 over a period τ_2, and maintained at that temperature for a period τ_1 (see Figure A.5.1).

In order to determine the dissipation function, we need to look at a time-symmetric protocol. For simplicity, we make the changes in T such that β varies linearly in time. To ensure ergodic consistency, we consider a Nosé–Hoover thermostatted system. This example can then be used to consider thermodynamically reversible or irreversible changes.

The equations of motion are

$$
\begin{aligned}
\dot{\mathbf{q}}_i &= \mathbf{p}_i/m \\
\dot{\mathbf{p}}_i &= \mathbf{F}_i - \alpha\mathbf{p}_i \\
\dot{\alpha} &= \frac{1}{\tau_{\text{th}}^2}\left(\frac{\mathbf{p}_i\cdot\mathbf{p}_i}{3NkT(t)\,m} - 1\right)
\end{aligned}
\tag{5.A1}
$$

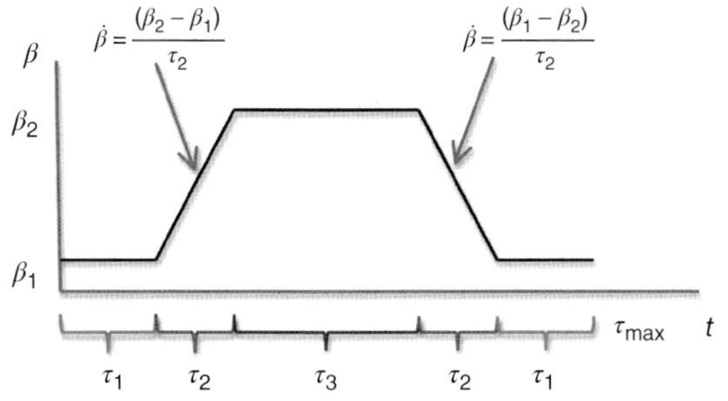

Figure A.5.1 Schematic diagram of the protocol used for change of temperature in the example considered. Reproduced from Williams, Searles, and Evans (2014) with permission of Taylor and Francis.

and the initial distribution function is

$$f(\mathbf{\Gamma}, \alpha) = \frac{e^{-\beta_1 H_0(\mathbf{\Gamma})-(3/2)N\tau_{\text{th}}^2\alpha^2}}{Z_1} \tag{5.A2}$$

This becomes a cyclic process if the time period τ_1 becomes long enough such that, for averages of low-order phase functions, the system approaches equilibrium. We will consider both possibilities here (cyclic and not).

The dissipation function for this process is

$$\Omega_{\tau_{\max}} = \beta_1 H_0(\tau_{\max}) - \beta_1 H_0(0) + \frac{3}{2}N\tau_{\text{th}}^2(\alpha(\tau_{\max})^2 - \alpha(0)^2) + 3N\int_0^{\tau_{\max}} \alpha(t)dt \tag{5.A3}$$

Noting that

$$\frac{d}{dt}\left[\frac{3}{2}N\tau_{\text{th}}^2\alpha(t)^2\right] = 3N\tau_{\text{th}}^2\alpha(t)\dot{\alpha}(t) = \frac{2K(t)\alpha(t)}{kT(t)} - 3N\alpha(t) \tag{5.A4}$$

so

$$\frac{3}{2}N\tau_{\text{th}}^2(\alpha(\tau_{\max})^2 - \alpha(0)^2) = \int_0^{\tau_{\max}} \left(\frac{2K\,(t)\,\alpha(t)}{k_B T(t)} - 3N\alpha(t)\right)dt \tag{5.A5}$$

and substituting into Eq. (5.A3) gives

$$\Omega_{\tau_{\max}} = \beta_1 H_0(\tau_{\max}) - \beta_1 H_0(0) + \int_0^{\tau_{\max}} \frac{2K(t)\alpha(t)}{k_B T(t)}dt \tag{5.A6}$$

Furthermore, $\dot{H}_0(t) = -2K(t)\alpha(t) = \dot{Q}(t)$, where $\dot{Q}(t)$ is the rate at which heat is transferred to the system, since no work is being done on the system. So

$$\Omega_{\tau_{\max}} = \beta_1 H_0(\tau_{\max}) - \beta_1 H_0(0) - \int_0^{\tau_{\max}} \frac{\dot{Q}(t)}{k_B T(t)}dt \tag{5.A7}$$

Now consider some special cases:

1) When $\lim(\tau_2 \to \infty)$, we have a reversible process. Then

$$\int_{\tau_1}^{\tau_1+\tau_2} \frac{\langle \dot{Q}\,(t)\rangle}{k_B T(t)}dt = -\int_{\tau_1+\tau_2+\tau_3}^{\tau_1+2\tau_2+\tau_3} \frac{\langle \dot{Q}\,(t)\rangle}{k_B T(t)}dt \quad \text{and} \quad \langle H_0\,(\tau_{\max})\rangle = \langle H_0\,(0)\rangle$$

so from (5.A7)

$$\left\langle \Omega_{\tau_{\max}} \right\rangle = -\int_0^{\tau_{\max}} \frac{\left\langle \dot{Q}(t) \right\rangle}{k_B T(t)} dt = 0 \tag{5.A8}$$

2) Now consider the irreversible process with finite τ_2 but with $\lim(\tau_1 \to \infty)$.

With respect to averages of low-order phase functions, the system will be arbitrarily close to equilibrium at $\tau_{\max}$, so $\lim_{\tau_{\max}\to\infty} \left\langle H_0\left(\tau_{\max}\right)\right\rangle = \left\langle H_0(0)\right\rangle$. Then, from Eq. (5.A7)

$$\left\langle \Omega_{\tau_{\max}} \right\rangle = -\int_0^{\tau_{\max}} \frac{\left\langle \dot{Q}(t) \right\rangle}{k_B T(t)} dt = -1/k_B \int_0^{\tau_{\max}} \dot{S}_G(t) dt \tag{5.A9}$$

and from the second law inequality (3.9), $\left\langle \Omega_{\tau_{\max}} \right\rangle > 0$, so

$$\left\langle \Omega_{\tau_{\max}} \right\rangle = -\int_0^{\tau_{\max}} \frac{\left\langle \dot{Q}(t) \right\rangle}{k_B T(t)} dt = -1/k_B \int_0^{\tau_{\max}} \dot{S}_G(t) dt > 0 \tag{5.A10}$$

If the process is irreversible, the inequality applies. The equality will apply for the reversible case. So this says that for the irreversible cycle, the time integral of the average dissipation function (multiplied by k_B), the change in the Gibbs entropy, and the integral of $\left\langle \dot{Q} \right\rangle / T$, where the temperature is the target temperature of the Nosé–Hoover thermostat, are all equal and will be positive, independent of the Nosé–Hoover time constant τ_{th}.

The target temperature will, in general, be different from the instantaneous kinetic temperature and, furthermore, those differences will vary with respect to the time constant τ_{th}. Exactly the same equation can be derived using an isokinetic rather than Nosé–Hoover thermostat. These facts show that the temperature $T(t)$ in Eq. (5.A10) is, in fact, the equilibrium thermodynamic temperature of the underlying equilibrium system at time t. This temperature can be discovered by halting the execution of the protocol at time t and allowing the entire system to relax to equilibrium. From the equilibrium relaxation theorems, for isokinetic dynamics this temperature is the instantaneous kinetic temperature at time t. For the Nosé–Hoover thermostat, it is the Nosé–Hoover target temperature at time t, regardless of the value of the feedback time constant.

Equation (5.A10) also shows the lack of utility of the Gibbs entropy in this work. Although its time derivative is $\left\langle \dot{Q}(t) \right\rangle / T(t)$, the difference in the Gibbs entropy of the initial and final states is not zero. This is in spite of the fact that an unlimited amount of time is allowed for relaxation toward the final state! For any relaxation time no matter how large, the final distribution at time $\tau_{\max}$ is not precisely an equilibrium distribution, and the Gibbs entropy detects these minute differences and $S_G(0) > S_G(\tau_{\max})$, $\forall \tau_{\max}$. If it did relax to a true equilibrium, we could never retrieve the initial distribution of states by applying a time-reversal operator. For any $\tau_{\max}$ no matter how large, the initial distribution of states can always be retrieved using a time-reversal operator. True equilibrium distributions are invariant in time with or without the application of time-reversal operators.

References

Berne, B.J., Boon, J.P., and Rice, S.A. (1966) On the calculation of autocorrelation functions of dynamical variables. *J. Chem. Phys.*, **45**, 1086–1096.

Boltzmann, L. (1872) Further studies on thermal equilibrium among gas molecules. *Akad. Wiss. Wien*, **66**, 275.

Butler, B.D., Ayton, O., Jepps, O.G., and Evans, D.J. (1998) Configurational temperature: verification of Monte Carlo simulations. *J. Chem. Phys.*, **109**, 6519–6522.

Cohen, E.G.D. (1990) George E. Uhlenbeck and statistical mechanics. *Am. J. Phys*, **56**, 618–625.

Daivis, P.J. and Todd, B.D. (2006) A simple, direct derivation and proof of the validity of the SLLOD equations of motion for generalized homogeneous flows. *J. Chem. Phys.*, **124**, 194103.

Dorfman, J.R. (1999) *An Introduction to Chaos in Nonequilibrium Statistical Mechanics*, Cambridge University Press, Cambridge.

Evans, D.J. (1981) Equilibrium fluctuation expressions for the wave-vector-dependent and frequency-dependent shear viscosity. *Phys. Rev. A*, **23**, 2622–2626.

Evans, D.J. (1983) Equilibrium fluctuation expressions for the wavevector and frequency dependent shear viscosity [*Phys. Rev. A*, **23**, 2622 (1981)], Erratum.. *Phys. Rev. A*, **27**, 1207.

Evans, D.J., Hoover, W.G., Failor, B.H., Moran, B., and Ladd, A.J.C. (1983) Nonequilibrium molecular-dynamics via gauss principle of least constraint. *Phys. Rev. A*, **28**, 1016–1021.

Evans, D.J. and Morriss, G.P. (1983) The isothermal isobaric molecular-dynamics ensemble. *Phys. Lett. A*, **98**, 433–436.

Evans, D.J. and Morriss, G.P. (1990) *Statistical Mechanics of Nonequilibrium Liquids*, Academic Press, London.

Evans, D.J., Searles, D.J., and Williams, S.R. (2009a) Dissipation and the relaxation to equilibrium. *J. Stat. Mech. -Theor. Exp.*, P07029.

Evans, D.J., Searles, D.J., and Williams, S.R. (2009b) A simple mathematical proof of Boltzmann's equal a priori probability hypothesis, in *Diffusion Fundamentals III* (eds C. Chmelik, N. Kanellopoulos, J. Karger, and T. Doros), Leipziger Universitatsverlag, Leipzig.

Evans, D.J., Williams, S.R., and Rondoni, L. (2012) A mathematical proof of the zeroth "law" of thermodynamics and the nonlinear Fourier "law" for heat flow. *J. Chem. Phys.*, **137**, 194109.

Evans, D.J., Williams, S.R., Rondoni, L., and Searles, D.J. (2016) Ergodicity of non-Hamiltonian equilibrium systems. arXiv:1602.06065 [cond-mat.stat-mech].

Green, H. (1952) *The Molecular Theory of Fluids*, Amsterdam, North-Holland.

Hoover, W.G. (1999) *Time Reversibility, Computer Simulation, and Chaos*, World Scientific, Singapore.

Huang, K. (1963) *Statistical Mechanics*, John Wiley & Sons, Inc, New York.

Irving, J.H. and Kirkwood, J.G. (1950) The statistical mechanical theory of transport processes. IV. The equations of hydrodynamics. *J. Chem. Phys.*, **18**, 817.

Lebowitz, J.L. (1993) Boltzmann's entropy and time's arrow. *Phys. Today*, **46**, 32–38.

Reid, J.C., Williams, S.R., Searles, D.J., Rondoni, L., and Evans, D.J. (2013) Fluctuation relations and the foundations of statistical thermodynamics: a deterministic approach and numerical demonstration, in *Nonequilibrium Statistical Physics of Small Systems* (eds R. Klages, W. Just, and C. Jarzynski), Wiley-VCH Verlag GmbH, Weinheim.

Sinai, Y.G. (1976) *Introduction to Ergodic Theory*, Princeton University Press, Princeton, NJ.

Thomson, W. (1874) Kinetic theory of the dissipation of energy. *Nature*, **9**, 441.

Thomson, C.J. (1972) *Mathematical Statistical Mechanics*, Collier Macmillan Ltd, London.

Tolman, R.C. (1979) *The Principles of Statistical Mechanics*, Dover, New York.

Williams, S.R. and Evans, D.J. (2007) Statistical mechanics of time independent nondissipative nonequilibrium states. *J. Chem. Phys.*, **127**, 184101.

Williams, S.R. and Evans, D.J. (2008) Statistical mechanics of time independent non-dissipative nonequilibrium states, in *Complex Systems, 5th International Workshop on Complex Systems* (eds

M. Tokyama, I. Oppenheim, and H. Nishiyama), American Institute of Physics.

Williams, S.R. and Evans, D.J. (2010) The rheology of solid glass. *J. Chem. Phys.*, **132**, 184105.

Williams, S.R., Searles, D.J., and Evans, D.J. (2008) The glass transition and the Jarzynski equality. *J. Chem. Phys.*, **129**, 134504.

Williams, S.R., Searles, D.J., and Evans, D.J. (2014) On the relationship between dissipation and the rate of spontaneous entropy production from linear irreversible thermodynamics. *Mol. Simul.*, **40**, 208–217.

Zwanzig, R. (1963) Elementary derivation of time-correlation formulas for transport coefficients. *J. Chem. Phys.*, **40**, 2527–2533.

6
Nonequilibrium Steady States

> The number of molecules being finite, it is clear that small finite deviations from absolute precision in the reversal we have supposed would not obviate the resulting disequalisation of the distribution of energy. But the greater the number of molecules, the shorter will be the time during which the disequalising will continue; and it is only when we regard the number of molecules as practically infinite that we can regard spontaneous disequalisation as practically impossible.[1]
>
> *(Thomson, 1874)*

6.1
The Physically Ergodic Nonequilibrium Steady State

Definition

As we saw in Section 2.2, a nonequilibrium system is *stationary* if it is subject to a thermostatting mechanism and a dissipative field such that for physical phase functions $A(\mathbf{\Gamma})$

$$\langle A(t)\rangle = \int_D d\mathbf{\Gamma} A(\mathbf{\Gamma})f(\mathbf{\Gamma};t) = \int_D d\mathbf{\Gamma} A(S^t\mathbf{\Gamma})f(\mathbf{\Gamma};0) = const, \ \ \forall t \tag{6.1}$$

where D denotes the ostensible phase space domain over which the initial ensemble density is nonzero. Stationarity simply means that such systems have time-independent averages for physical properties. This stationarity may occur for all times or only for sufficiently late times. Stationarity does not imply that the distribution function is stationary. In fact, as we first saw in Eq. (2.56), in a nonequilibrium stationary state the Gibbs entropy diverges at a constant rate toward negative infinity. In stationary nonequilibrium states, the time-independent value that these time averages take on can be dependent on the initial phase $\mathbf{\Gamma}$.

1) In modern terminology, Thomson is saying, other things being equal, that the largest Lyapunov exponent increases with the number of particles, so that its reciprocal (the reversal time) becomes microscopic for macroscopic systems. At the present time it is unknown whether Thomson's assertion is correct: (Searles, Evans, and Isbister, 1997).

Fundamentals of Classical Statistical Thermodynamics: Dissipation, Relaxation and Fluctuation Theorems, First Edition. Denis J. Evans, Debra J. Searles, and Stephen R. Williams.

Definition

A *physically ergodic nonequilibrium steady state* (*peNESS)* satisfies the equation

$$\lim_{t\to\infty} \langle A(t)\rangle_0 = \lim_{t\to\infty} \overline{A}_t(\mathbf{\Gamma}) = \lim_{t\to\infty} \frac{1}{t}\int_0^t ds\ A(S^s\mathbf{\Gamma}), \quad \forall!\mathbf{\Gamma} \in D \tag{6.2}$$

where $\langle\cdots\rangle_0$ denotes an ensemble average over the *initial* $t = 0$ and ensemble $f(\mathbf{\Gamma}; 0)$. For almost any ∀! initial phase $\mathbf{\Gamma}$, the *time average* on the left-hand side of Eq. (6.2) equals the right-hand side *late-time ensemble average* taken with respect to the initial distribution $f(\mathbf{\Gamma}; 0)$ in D. At late times, we say that the steady state is ergodic with respect to ensemble averages of physical properties over the initial distribution in D.

When we speak of physical ergodicity, we say that *almos*t any initial phase leads, at long times, to time averages that are equal to the long-time ensemble average, because there may be sets of initial phase points, with zero measure, whose long-time averages of physical properties are different from those of almost all other initial phases (e.g., the zero measure phases that generate the antisteady state).

Definition

By *zero measure* we mean that, if we sample phases randomly over the ostensible phase space domain D, the probability of observing these zero measure phases is *precisely* zero.

Our notion of physical ergodicity is defined solely by averages of physical phase variables. We make no statement about the nonequilibrium distributions $f(\mathbf{\Gamma}; t)$. These distributions are for macroscopic systems, are practically impossible to measure, and they frequently relax toward fractal objects.

Consider almost any $\mathbf{\Gamma} \in D$ if we form the time-reversal mapping of the set of late-time phase vectors $\{S^t\mathbf{\Gamma}\}$; then the set of phases $\{M^T S^t\mathbf{\Gamma}\}$ has very strange properties. Advancing time for a short while $\tau \ll t$ generates an antisteady state ensemble in which it is overwhelmingly likely to convert heat into work! However, for sufficiently large t, the probability of observing these "repeller" phases in a typical set of equilibrium phases $\{\mathbf{\Gamma}\}$ becomes incredibly small – going to zero in the limit $t \to \infty$. This is why we use the terminology "almost any" initial phase will generate a trajectory along which the time-averaged dissipation is positive. This is the qualitative content of the transient fluctuation theorem.

Equation (6.2) implies that the long-time averages could be averaged over almost any initial distribution of those initial phases $f(\mathbf{\Gamma}; 0)$ so that for a peNESS

$$\lim_{t\to\infty} \left\langle B(S^t\mathbf{\Gamma}\right\rangle_0 \equiv \lim_{t\to\infty} \int_{\mathbf{\Gamma}\in D} d\mathbf{\Gamma}\, B(S^t\mathbf{\Gamma})f(\mathbf{\Gamma}; 0) = \lim_{t\to\infty} \left\langle B(S^t\mathbf{\Gamma}\right\rangle \quad \forall! f(\mathbf{\Gamma}; 0), \ \ \forall\mathbf{\Gamma} \in D \tag{6.3}$$

Equation (6.3) shows that peNESSs are physically ergodic with respect to almost *any* initial distribution of phases $f(\mathbf{\Gamma}; 0)$ in the phase space domain D. Time averages of physical variables are equal to ensemble averages, and in the nonequilibrium steady state (NESS) the late-time ensemble average is independent of the initial distribution. This coincides with the commonplace observation that, for example, in shearing laminar flows with fixed boundary conditions of shear rate, boundary temperatures and pressure, and so on, the NESS so produced is, at long times, independent of the details of the initial state from which it was produced and a single late-time time average will give the same result for a physical measurement that we would obtain from a late-time ensemble average of repeated experiments.

Later in this chapter we will examine in more detail the mathematical conditions that are required in order to generate a peNESS.

While our physical experience is that physically ergodic NESSs do exist, we also know from experience that NESSs are not always physically ergodic. For example, in Rayleigh–Bénard instability we know that for a fixed geometry and a given set of boundary conditions, systems may form into two, or four rolls, and so on, with the number of rolls fixed and persisting (apparently) indefinitely. Clearly, the physical properties of the two-roll system are different from those of a four-roll system. These types of system do not satisfy Eq. (6.2) (i.e., they do not form a peNESS).

If we could define the initial phase space subdomain D_2 over which *only* two rolls form at late times, then over the subdomain D_2 the system would form a peNESS, while over D it would not. However, in such cases it may be practically impossible to actually discover the topology of this phase space subdomain D_2. Such domains are expected to be fractal. If this is the case, then from a practical point of view such systems are best viewed as not forming a physically ergodic NESS.

6.2 Dissipation in Nonequilibrium Steady States (NESSs) (Williams, Searles, and Evans, 2014)

We will now discuss dissipation (rather than entropy or the so-called entropy production) in NESSs. We begin by considering a system of N particles subject to the following equations of motion:

$$\dot{\mathbf{q}}_i = \mathbf{p}_i/m + C_i\mathbf{F}_e, \quad \dot{\mathbf{p}}_i = \mathbf{F}_i + D_i\mathbf{F}_e - S_i\alpha_{IK}\mathbf{p}_i + S_i\mathbf{F}_{\text{th}} \tag{6.4}$$

In these equations, $\mathbf{F}_e$ is an external dissipative field (e.g., an electric field applied to a molten salt), and the scalar phase functions C_i and D_i couple the system to the field. The system can be easily generalized to tensor coupling parameters if required. If we denote a set of thermostatted particles as belonging to the set th, we choose $S_i = 0,\ \ i \notin \text{th}; = 1, i \in \text{th}$ to be a switch to determine whether particle i is a member of the set, *th*, of N_{th} thermostatted particles. α_{IK} is the thermostat multiplier chosen to fix the kinetic energy of the thermostatted particles at the value K_{th}, and $\mathbf{F}_{\text{th}}$ is a fluctuating force used to fix the total momentum of the thermostatted particles, which is chosen to have a value of zero. We assume the interatomic forces $\mathbf{F}_i;\ \ i = 1, N$, are smooth functions of the interparticle separation. We also assume that the interatomic forces are short-ranged so that there are no convergence problems in the large N limit.

We assume that in the absence of the thermostatting and momentum-zeroing forces, the equations of motion preserve phase space volumes

($\partial/\partial\Gamma \cdot \dot{\Gamma}^{ad} \equiv \Lambda^{ad}(\Gamma) = 0$) where $\Gamma \equiv (\mathbf{q}_1, \ldots, \mathbf{q}_N, \mathbf{p}_1, \ldots, \mathbf{p}_N)$ is the phase space vector and the superscript "ad" denotes the fact that the time derivative is calculated with the thermostatting and momentum zeroing forces turned off. This condition is known as the *adiabatic incompressibility of phase space condition* or $AI\Gamma$ for short – see Section 2.2.

We assume that the system of particles is subject to infinite checkerboard boundary conditions (Evans and Morriss, 1990) – at least in the direction of the force. This means that angular momentum is not a constant of the motion. It also means that dissipation can go on forever without the system relaxing to equilibrium. Currents can flow in the direction of the force forever. The thermostatted particles may be taken to form solid walls parallel to the field, so that they can absorb or liberate heat that may be required to generate a NESS characterized by a fixed value for the kinetic energy of the thermostatted particles.

In contrast, if the system is finite, mixing, and has an autonomous Hamiltonian, even when subject to a *dissipative* external force it will eventually relax toward microcanonical equilibrium (Section 5.3). If these same systems are thermostatted, as in Eq. (6.4), they will eventually relax toward canonical equilibrium (Section 5.4). For example, a finite cell containing charged particles subject to a fixed external field, whether thermostatted or not, will eventually, after dissipative transients, relax toward equilibrium. The charges will be separated by the external field and eventually produce an internal field (space charge) that cancels out the externally applied field.

However, although NESSs that persist for an infinite amount of time do not exist in Nature, on accessible timescales they can be approached arbitrarily closely by a judicious choice of large but finite heat reservoirs and managing the magnitude of dissipation in relation to the size of those reservoirs and the nonequilibrium system of interest. If the time taken to relax toward equilibrium is much longer than the time taken to relax toward a (transient) nonequilibrium "steady" state, averages of low order phase functions in those transient dissipative states can be approximated as stationary averages.

In this chapter we consider only those particles that are initially located in the unit cell at time zero. The equations of motion given in Eq. (6.4) do not need to refer to the periodic boundaries or re-imaging processes because we follow the coordinates on this initial set of particles indefinitely no matter how far they may diffuse or stream from the initial unit cell, and the coordinates of the particles are then continuous in time. No matter where one of the original particles is located at later times, the force on that particle due to any one of the infinite periodic array of other particles close enough to exert a force on this original particle is computed correctly. This is done by exploiting the infinite checkerboard convention. At long times, the nearest neighbors of one of the original unit cell particles are not necessary members of the original unit cell. This is the so-called infinite checkerboard convention commonly used in molecular dynamics and Monte Carlo computer simulation (Allen and Tildesley, 1987).

The initial distribution is taken to be the equilibrium distribution for this system (see below). It takes the form of a canonical phase space distribution function $f_c(\Gamma)$

augmented with the necessary delta functions (5.12):

$$f(\mathbf{\Gamma};0) = f_c(\mathbf{\Gamma}) = \frac{\exp[-\beta_{\text{th}}H_0(\mathbf{\Gamma})]\delta(\mathbf{P}_{\text{th}})\delta(K_{\text{th}}(\mathbf{\Gamma}) - K_{\beta,\text{th}})}{\int d\mathbf{\Gamma}\, \exp[-\beta_{\text{th}}H_0(\mathbf{\Gamma})]\delta(\mathbf{P}_{\text{th}})\delta(K_{\text{th}}(\mathbf{\Gamma}) - K_{\beta,\text{th}})} \tag{6.5}$$

where $\mathbf{P}_{\text{th}} = \sum_{i=1}^{N} S_i\mathbf{p}_i$ is the total momentum of the thermostatted particles, $K_{\text{th}}(\mathbf{\Gamma}) = K_{\text{th}}(p) = \sum S_i p_i^2/2m_i$ is the kinetic energy of the thermostatted particles, and $K_{\beta,\text{th}} = (3N_{\text{th}} - 4)\beta_{\text{th}}^{-1}/2$ (we assume the system has three Cartesian dimensions) is the fixed *value* of the kinetic energy of the thermostatted particles. The number of particles in a unit cell is N. The kinetic energy of the thermostatted particles is fixed using the Gaussian multiplier α_{IK} in the equations of motion. Here, $\beta_{\text{th}} = 1/k_B T_{\text{th}}$, where k_B is Boltzmann's constant and, for isokinetic systems, T_{th} is the so-called kinetic temperature of the thermostatted particles. For Nosé–Hoover thermostatted systems, it is the reciprocal of the *target* temperature of the Nosé–Hoover feedback mechanism. In the Nosé–Hoover thermostatted case, there is an $O(1)$ change in the equipartition relation between the thermostat kinetic energy and the kinetic temperature of the thermostat (Evans and Morriss, 1990). The (only) common feature of *all* thermostatted systems is that β_{th} is the reciprocal of the equilibrium thermodynamic temperature that the entire driven system would relax toward if the system is T-mixing Eq. (5.10), the driving force is set to zero, and the whole system is allowed time to relax toward thermodynamic equilibrium under the influence of the thermostat (Chapter 5). We call this temperature the thermodynamic temperature of the *underlying* equilibrium state. The internal energy of the N particles in the unit cell is the average of $H_0(\mathbf{\Gamma}) = K(\mathbf{p}) + \Phi(\mathbf{q})$, where K, Φ are, respectively, the peculiar kinetic energy and the potential energy of all the particles in the original unit cell.

To be more mathematically correct, we should specify the ostensible phase space domain that is not referred to explicitly in Eq. (6.5). In principle, the particle momenta are unbounded. Clearly, the delta functions in Eq. (6.5) place four constraints on the momenta of (some) particles in the system. The initial coordinates of the particles will each range over some finite range $\pm L$ within the unit cell of the periodic system. Because of the infinite periodicity, any particle and its environment are identical to any periodic image of that particle. Particles can always be "re-imaged" back into the original unit cell (Evans and Morriss, 1990). However, calculating certain quantities may have spurious discontinuities if this is done. Thermodynamic quantities such as pressure, internal energy, and so on, are all continuous in time, independent of whether particles are "imaged" in the unit cell. Throughout most of the remainder of this paper, we will not refer explicitly to this ostensible phase space domain.

The thermostatting region that is unnatural can be made arbitrarily remote from the natural system of interest. The thermostatting particles may be buried far inside realistically modeled walls that contain the nonequilibrium system of interest. This means that there is no way that the particles in the system of interest can "know" how heat is ultimately being removed from the system. The thermostats are important as a bookkeeping device to track the evolution of phase space volume in a deterministic but open system.

The time integral of the dissipation function evaluated at an initial phase $\mathbf{\Gamma}$ is formally defined as

$$\Omega_t(\mathbf{\Gamma}) \equiv \ln \left[\frac{f(\mathbf{\Gamma};0)\exp\left(-\int_0^t ds\,\Lambda(S^s\mathbf{\Gamma})\right)}{f(M^T S^t\mathbf{\Gamma};0)} \right] \tag{6.6}$$

where M^T is the time reversal map, $M^T\mathbf{\Gamma} \equiv (\mathbf{q}_1, \ldots, \mathbf{q}_N, -\mathbf{p}_1, \ldots, -\mathbf{p}_N)$, and S^t is the time evolution operator for a time t. A key point in the definition of dissipation is that $\mathbf{\Gamma}$ and $M^T S^t\mathbf{\Gamma}$ are the origin phases for a trajectory and its conjugate antitrajectory, respectively. This places constraints on the propagator S^t. Any time-dependent driving fields $\mathbf{F}_e(t)$ *must* have a definite parity under time reversal over the interval $(0, t)$.

For a system satisfying Eq. (6.4) and satisfying the AI$\mathbf{\Gamma}$ condition and having an initially equilibrium distribution of states Eq. (6.5), it is easy to show that the instantaneous dissipation function (6.6) can be written as

$$\Omega(\mathbf{\Gamma}) \equiv -\beta_{\text{th}} V\mathbf{J}(\mathbf{\Gamma}) \cdot \mathbf{F}_e = \beta_{\text{th}} \sum_i [\mathbf{p}_i D_i/m - \mathbf{F}_i C_i] \cdot \mathbf{F}_e \tag{6.7}$$

where $\mathbf{J}(\mathbf{\Gamma})$ is the so-called dissipative flux and V is the unit cell volume. For example, for electrical conductivity, where $C_i = 0, \forall i$ and $D_i = c_i$ is the electric charge of particle i and an electric field is applied in the x-direction, $\mathbf{F}_e = (F_e, 0, 0)$, it is easy to see that $-JV = \sum c_i \dot{x}_i$, the electric current in the x-direction.

Such a dissipation function is called a *primary dissipation function* – Section 4.3. When the field is zero, the system remains in equilibrium and the dissipation is identically zero.

From the dissipation theorem (Chapter 4) we know that, if the system is initially at equilibrium, we can write the nonlinear response of an arbitrary integrable phase function $B(\mathbf{\Gamma})$ as

$$\langle B(t)\rangle_{F_e,0} = \langle B(0)\rangle_0 - \beta_{\text{th}} V \int_0^t ds \langle \mathbf{J}(0)B(s)\rangle_{F_e,0} \cdot \mathbf{F}_e \tag{6.8}$$

where $\langle B(t)\rangle_{F_e,0}$ denotes the ensemble average of the phase function $B(\mathbf{\Gamma})$ evaluated at the propagated phase $S^t\mathbf{\Gamma}$, with the initial distribution ($t = 0$) being given by Eq. (6.5). The first subscript on the ensemble averages F_e indicates that the propagator S^t is given by the full field-dependent thermostatted dynamics of Eq. (6.4), and the second subscript, which is zero in this case, indicates that the average is with respect to the initial equilibrium distribution function. In Eq. (6.8), $B(s) \equiv B(S^s\mathbf{\Gamma})$, $\mathbf{J}(0) \equiv \mathbf{J}(\mathbf{\Gamma})$, and $B(s)$ is also evaluated with the full field-dependent thermostatted dynamics.

From Eq. (6.8) we also see that, if the system is initially at equilibrium and the driving field is zero, then the ensemble averages of *all* integrable phase functions are time independent, and if the system starts with the equilibrium distribution (6.5), the distribution is preserved by the field-free thermostatted dynamics.

Although Eq. (6.8) only refers directly to the N particles in the unit cell, the coordinates and momenta of all periodic image particles follow by symmetry.

This expression (6.8) is exact for systems arbitrarily near or far from equilibrium and also for systems of arbitrary size. If the system is ΩT-mixing, then by *definition* (5.10), if $B(\mathbf{\Gamma})$ is a real-valued physical phase function

$$\lim_{t\to\infty} \langle B(t)\rangle_{F_e,0} = \langle B(0)\rangle_{F_e,0} + L_0 \in \Re \tag{6.9}$$

So the infinite time integral of the transient time correlation function in Eq. (6.8) converges to a finite value for an arbitrarily strong, or weak, dissipative field. Physically this corresponds to the relaxation to a NESS at sufficiently long times. The T-mixing property Eq. (5.11) also means that the steady-state distribution is physically ergodic and does not break down into nonmixing subspaces that have different values for the steady-state averages of low-order phase functions.

If the phase space did break down into subdomains with distinct time-invariant sets of averages of these low-order phase functions, these distinct values could be used to define new constants of the motion. These constants of the motion would lead to nonconvergent integrals for the relevant time correlation functions, and thereby violate the T-mixing assumption. Any initial correlations between the physical phase functions that are constants of the motion would be preserved for all time, resulting in divergences in Eq. (6.8).

Thus steady-state time averages must equal ensemble averages over the steady state attractor(s) even though its topology is fractal and its geometry is generally unknown. So the T-mixing condition implies that the late-time stationary nonequilibrium states are in fact NESSs that are physically ergodic over the initial phase space domain.

We do know that the dimension of the steady-state attractor(s) is less than that of the ostensible phase space and generally decreases as the dissipation increases. (N.B. it is not known whether the Kaplan–Yorke dimension is in general a monotonic decreasing function of the dissipative field. In the weak field, linear response regime the Kaplan–Yorke dimension is a monotonic decreasing function of the dissipative field (Evans *et al.*, 2000).)

Although the long-time averages appearing on the left-hand side of Eqs. (6.8) and (6.9) are finite, these averages could be divergent in the limit of large system size. For finite dissipative fields, the large system limit is usually problematic. For example, for a fixed shear rate in Couette flow, as the system size increases, so does the Reynolds number for the flow. As the system size increases, we know that there will be a transition from laminar to (eventually) highly turbulent flow. Such large systems (e.g., Rayleigh–Bénard flows) may not be T-mixing at high fields, and once it becomes turbulent, it is clearly not T-mixing as a steady state is not achieved.

In the weak field limit, this Eq. (6.8) reduces (essentially) to the well-known Green–Kubo expression (Evans and Morriss, 1990; McQuarrie, 1976) for the linear response:

$$\lim_{F_e\to 0} \langle B(t)\rangle_{F_e,0} = \langle B(0)\rangle_0 - \beta_{\text{th}} V \int_0^t ds\langle \mathbf{J}(0)B(s)]\rangle_{F_e=0,0} \cdot \mathbf{F}_e \tag{6.10}$$

where the right-hand side is given by the integral of an *equilibrium* (i.e., $\mathbf{F}_e = \mathbf{0}$) time correlation function. The initial ensemble for the terms on the right-hand side is the equilibrium ensemble Eq. (6.5), and the dynamics inherent in the equilibrium time correlation function is generated at zero field but with the thermostat on. The field only appears in the nonequilibrium average on the left-hand side of Eq. (6.10) and as an explicit factor multiplying the correlation function on the right-hand side of Eq. (6.10).

In the linear response regime, the T-mixing condition implies that, for a system initially in a canonical ensemble, $\lim_{t\to\infty}\langle J(0)B(t)\rangle_{F_e=0,C} = 0$; that is, in the long-time limit there is no correlation between $J(0)$ and $\lim_{t\to\infty}B(t)$ so that $\lim_{t\to\infty}\langle J(0)B(t)\rangle_{F_e=0,C} = \langle J\rangle_{F_e=0,C}\langle B\rangle_{F_e=0,C} = 0$ and the zero-field system is *mixing*. If correlations were nonzero in the long-time limit, the integral Eq. (6.10) could not converge and the system would not be T-mixing. The mixing condition assumes that Eq. (5.1)

$$\lim_{t\to\infty}\left|\langle A(\mathbf{\Gamma}(0))\,B(\mathbf{\Gamma}(t))]\rangle_{F_e=0,C} - \langle A(\mathbf{\Gamma})\rangle_{F_e=0,C}\langle B(\mathbf{\Gamma})\rangle_{F_e=0,C}\right| = 0 \tag{6.11}$$

where the subscript, C, on the ensemble average implies it is a canonical ensemble average and $f_c(\mathbf{\Gamma})$ is the equilibrium distribution (6.5). In Eq. (6.8), the T-mixing condition for the equilibrium time correlation function implies that the system is mixing over the invariant equilibrium distribution (6.5). In the zero-field limit, the T-mixing transient time correlation function must be a rapidly decaying mixing equilibrium time correlation function.

However, mixing does not, in general, imply T-mixing. In a T-mixing system, the correlation function must go to zero sufficiently rapidly for the integral to converge, so that $\lim_{t\to\infty}|\langle B(\mathbf{\Gamma}(t))\rangle_{F_e,0}|$ is time-independent and *finite.* Having equilibrium time correlation functions going to zero at long times is insufficient to ensure T-mixing.

If the decorrelations in the equilibrium time correlation function scale like $1/t$ or slower at long times, the system will be mixing but not T-mixing. For example, if the equilibrium time correlation function goes to zero as $1/t$ at long times, $\lim_{t\to\infty}\lim_{F_e\to 0}\langle B(t)\rangle_{F_e} - \langle B(0)\rangle = O(\ln(t))$ and the system will *never* have a time-independent average value for the phase variable even arbitrarily close to equilibrium. Thus we have an example system that could be mixing over the equilibrium time correlation function but does not relax to NESSs in the linear response regime close to equilibrium. This is quite different to the ergodic theory result for finite, autonomous Hamiltonian systems where a mixing equilibrium distribution that is preserved by the dynamics does indeed imply relaxation toward that time-independent microcanonical equilibrium distribution! For such systems, the transients must in fact be at least ΩT-mixing.

To put this into a more physical context, in two dimensions in the large system limit (i.e., the number of particles N in the unit cell goes to infinity), equilibrium time correlation functions for the macroscopic Navier–Stokes transport coefficients are each thought to have t^{-1} long time tails (Alder and Wainwright, 1970). In this limit, the Fourier series in our infinitely periodic system become continuous Fourier–Laplace transforms. Thus *macroscopic* equilibrium systems in two dimensions may be mixing but would not be T-mixing. Again, this does not violate the ergodic theory proof of relaxation to microcanonical equilibrium because that proof applies only to finite systems.

If we turn briefly to the *transient* time correlation function expressions for the nonlinear response Eq. (6.8), the mixing condition is simply not relevant. The transient time correlation function on the right-hand side is not stationary. The measure evolves from the initial equilibrium distribution (6.5) and through a set of transient measures (over which the transient integral is computed) until at long times, if the system is ΩT-mixing, we approach the steady state with stationary averages for physical observables. The mixing condition (6.11) can never prove relaxation to a steady state because the condition *already* assumes in its definition (6.11) stationarity with respect to time!

We say that Eq. (6.10) is "essentially" the same as the Green–Kubo relations because there are some subtle differences. Kubo's results (Kubo, 1966) were for the linearized adiabatic response (i.e., no thermostats) of a canonical ensemble of systems. We derived Eq. (6.10) for isokinetic dynamics where the kinetic energy of the thermostatted particles is fixed and the distribution for the system of interest is canonical – Eq. (6.5). Thus the equilibrium time correlation function appearing in Eq. (6.10) is for field-free isokinetic dynamics. This is not the same as the case considered by Kubo. Kubo's time correlation functions involved canonical distributions but the field-free dynamics employs constant-energy Newtonian trajectories. Kubo's system was obviously not physically ergodic (because states of different energies never mix), whereas our results, Eq. (6.10), are physically ergodic (because the system is T-mixing). An equivalent result to Eq. (6.10) can be obtained for systems that are initially at equilibrium (with a canonical distribution function) but with thermostat-free dynamics for $t > 0$. The dissipation function for this system will be the same as in the thermostatted case. Then, Eq. (6.10) would be equivalent to the Green–Kubo relations.

Evans and Sarman have proved (Evans and Sarman, 1993) that, to leading order in the number of degrees of freedom in the system (= $O(N)$), adiabatic and homogeneously thermostatted equilibrium correlation functions are identical. Of course, if the dissipative field only couples to particles in the system of interest and the thermostat region is large and remote, the fluctuations in the dissipation function (which is local to the system of interest) will hardly be affected by the presence or absence of thermostatting terms in the large remote thermostatting region.

Because the thermostat is unphysical, we only thermostat a subset of the total number of particles. If we only thermostat particles that are remote from the natural system of interest (still within the unit cell), we can always appeal to the gedanken experiment that if we make the thermostatting region ever more remote from the system of interest, there is just no way that the physical system of interest can "know" how the remote thermostatting is actually occurring. If the external fields are set to zero and the T-mixing system is allowed to relax to equilibrium, we know the thermodynamic temperature of that underlying equilibrium system. That is the temperature that appears in the equations given above.

In fact, Evans and Sarman also proved (Evans and Sarman, 1993) that at the same state point, steady-state time correlation functions and steady-state averages computed for homogeneously isokinetic, isoenergetic, or Nosé–Hoover dynamics are identical to leading order in N. The proof assumes that the NESS is mixing. To define a common state point, they fixed N, V, F_e, and the average value of the thermostat multiplier $\langle \alpha \rangle$. It is easy to understand how this happens. If the system is T-mixing, it will relax to a mixing steady state. The ensemble-averaged value for the thermostat multiplier becomes constant in time. Since $\lim_{t\to\infty}\langle \alpha(t)\rangle = \text{const}$ independent of the thermostatting/ergostatting mechanism, in the large system limit the differences in the Liouvilleans and propagators are $O(1/N)$. In the large system limit, there is no practical difference between constant-energy or constant-kinetic-temperature thermostats and ergostats.

There is yet another interesting observation we can make regarding Kubo's system. If you consider viscous flow in a dilute gas, then as is known from kinetic theory, the viscosity of a gas increases with temperature (Huang, 1963). This means that for any finite field, no matter how small, the shear stress of an adiabatic shearing gas must increase with time. This means that a shearing unthermostatted gas can never be T-mixing! In a physical sense, for such a system the time correlations never decay – at least not rapidly enough for T-mixing.

You can see how this memory effect occurs. If among the initial ensemble members one encounters a fluctuation that increases the gas viscosity, that fluctuation will cause slightly more heating of the gas. In this slightly heated gas, the viscosity will be slightly higher than average, increasing the likelihood of further fluctuations that in turn increase the viscosity. This is a runaway process that prevents the decay of correlations required for the T-mixing condition.

6.3
For T-Mixing Systems, Nonequilibrium Steady-State Averages are Independent of the Initial Equilibrium Distribution

We have already argued that for T-mixing systems the steady-state properties must be independent of the initial distribution. In this section we give an explicit proof of this point (Evans *et al.*, 2015).

The dissipation function defined in Eq. (6.6) is a functional of *both* the dynamics and the initial distribution. The exact transient Evans–Searles fluctuation theorem refers to this exact dissipation function. How does the influence of the nonequilibrium initial distribution disappear? We consider an initial distribution that is *not* the equilibrium distribution for the zero-field system but is some deviation from it:

$$f_g(\boldsymbol{\Gamma};0) = \frac{\exp[-\beta_{\text{th}}H_0(\boldsymbol{\Gamma}) + \lambda g(\boldsymbol{\Gamma})]\delta(\mathbf{P}_{\text{th}})\delta(K_{\text{th}}(\boldsymbol{\Gamma}) - K_{\beta,\text{th}})}{\int d\boldsymbol{\Gamma}\, \exp[-\beta_{\text{th}}H_0(\boldsymbol{\Gamma}) + \lambda g(\boldsymbol{\Gamma})]\delta(\mathbf{P}_{\text{th}})\delta(K_{\text{th}}(\boldsymbol{\Gamma}) - K_{\beta,\text{th}})} \tag{6.12}$$

We assume that the deviation function $g(\boldsymbol{\Gamma})$ is even in the momenta, is nonsingular, real, and integrable, and the system is ergodically consistent. The positive real parameter λ is a simple scaling parameter $[0 \leq \lambda \leq 1]$ that allows us to easily scale the magnitude of the deviation from the equilibrium distribution. The dissipation function is easily seen to be

$$\Omega_\lambda(\boldsymbol{\Gamma}) = -\beta_{\text{th}}\mathbf{J}(\boldsymbol{\Gamma})V \cdot \mathbf{F}_e - \lambda\dot{g}(\boldsymbol{\Gamma}) \tag{6.13}$$

Substituting into the dissipation theorem gives

$$\langle g(t)\rangle_{\mathbf{F}_e,\lambda} = \langle g(0)\rangle_{\mathbf{F}_e,\lambda} - \int_0^t ds\langle[\beta_{\text{th}}\mathbf{J}(0)V \cdot \mathbf{F}_e + \lambda\dot{g}(0)]g(s)]\rangle_{\mathbf{F}_e,\lambda} \tag{6.14}$$

and recalling that $g(t)$ is even in the momenta we have $\langle\dot{g}(0)\rangle_{\mathbf{F}_e,\lambda} = \langle\mathbf{J}(0)\rangle_{\mathbf{F}_e,\lambda} = 0$. So, if the system is T-mixing, then at sufficiently long times the value of the left-hand side becomes time-independent, which means that the part of the ensemble-averaged dissipation function due specifically to the deviation function (6.14) is zero at long times:

$$\lim_{t\to\infty} \langle\dot{g}(t)\rangle_{\mathbf{F}_e,\lambda} = 0 \tag{6.15}$$

It is logically possible that at long times the ensemble-averaged dissipative flux $\lim_{t\to\infty} \langle\mathbf{J}(t)\rangle_{\mathbf{F}_e,\lambda} \neq \lim_{t\to\infty} \langle\mathbf{J}(t)\rangle_{\mathbf{F}_e,\lambda=0}$ is still influenced by the initial deviation function. For T-mixing systems, we now give a formal proof that this is not so and that at long times the dissipation becomes independent of the deviation function.

We write the average dissipation for the deviated system as

$$\begin{aligned}
\lim_{t\to\infty} \langle\Omega_\lambda(t)\rangle_{\mathbf{F}_e,\lambda} &= -\lim_{t\to\infty} \langle\beta_{\text{th}}\mathbf{J}(t)V \cdot \mathbf{F}_e + \lambda\dot{g}(t)\rangle_{\mathbf{F}_e,\lambda} \\
&= -\lim_{t\to\infty} \langle\beta_{\text{th}}\mathbf{J}(t)V \cdot \mathbf{F}_e\rangle_{\mathbf{F}_e,\lambda} \\
&= -\lim_{t\to\infty} \frac{\langle\beta_{\text{th}}\mathbf{J}(t)V \cdot \mathbf{F}_e e^{\lambda g(0)}\rangle_{\mathbf{F}_e,\lambda=0}}{\langle e^{\lambda g(0)}\rangle_{\mathbf{F}_e,\lambda=0}} \\
&= -\lim_{t\to\infty} \frac{\langle\beta_{\text{th}}\mathbf{J}(t)V \cdot \mathbf{F}_e\rangle_{\mathbf{F}_e,\lambda=0}\langle e^{\lambda g(0)}\rangle_{\mathbf{F}_e,\lambda=0}}{\langle e^{\lambda g(0)}\rangle_{\mathbf{F}_e,\lambda=0}} \\
&= -\lim_{t\to\infty} \langle\beta_{\text{th}}\mathbf{J}(t)V \cdot \mathbf{F}_e\rangle_{\mathbf{F}_e,\lambda=0} = \lim_{t\to\infty} \langle\Omega_{\lambda=0}(t)\rangle_{\mathbf{F}_e,\lambda=0}
\end{aligned} \tag{6.16}$$

In going from the first to the second line, we used Eq. (6.15). Going from the third line to the fourth, we used the fact that at long times the system becomes stationary and the T-mixing transient system must also be weak T-mixing – see Eq. (5.12). (Note: T-mixing implies weak T-mixing but weak T-mixing does not imply T-mixing.)

6.4
In the Linear Response Steady State, the Dissipation is Minimal with Respect to Variations of the Initial Distribution

In the linear response regime (where fluxes are linear in $\mathbf{F}_e$, λ), where the average dissipation function is quadratic in $\mathbf{F}_e$, λ, for finite times we have

$$\begin{aligned}\lim_{\mathbf{F}_e,\lambda\to 0,} \int_0^t ds \langle \Omega_\lambda(s)\rangle_{\mathbf{F}_e,\lambda} &= -\int_0^t ds \left[\beta_{\text{th}} V\mathbf{F}_e \cdot \lim_{\mathbf{F}_e\to 0} \langle \mathbf{J}(s)\rangle_{\mathbf{F}_e,\lambda=0} + \lambda \lim_{\lambda\to 0,} \langle \dot{g}(s)\rangle_{\mathbf{F}_e=\mathbf{0},\lambda}\right] \\ &> -\int_0^t ds\beta_{\text{th}} V\mathbf{F}_e \cdot \lim_{\mathbf{F}_e\to 0} \langle \mathbf{J}(s)\rangle_{\mathbf{F}_e,\lambda=0} = O(F_e^2) \\ &> 0, \quad \forall t \end{aligned} \tag{6.17}$$

It is easy to show that any cross-terms $\lambda\mathbf{F}_e \cdot \langle \mathbf{J}(0)\dot{g}(t)\rangle_{\mathbf{F}_e=0,\lambda=0}$ vanish by symmetry in the linear response regime, and in any case are of higher order $O(F_e^2\lambda^2)$. The second line in this equation follows from applying the second law inequality to the (weak) dissipation due solely to the deviation function: $-\int_0^t ds\ [\lim_{t\to\infty} \langle \dot{g}(s)\rangle_{\mathbf{F}_e=\mathbf{0},\lambda}] > 0, \quad \forall t$. The last line follows by applying the second law inequality to the dissipation due to both dissipative fields. We note that at any moment in time, the ensemble-averaged dissipation from either the dissipative field or the initial deviation function, or both, may be negative.

The second line of Eq. (6.17) shows that, in the linear response steady state, the time-averaged primary dissipation is less than any other dissipation due to variations in the initial distribution away from equilibrium. In the nonlinear regime, it is not known whether the average primary dissipation is minimal.

Equation (6.17) shows that at sufficiently long times, in T-mixing driven systems the dissipation always relaxes toward the average of the primary dissipation function. If the driven system is T-mixing, all other forms of dissipation decay toward zero, leaving only the primary dissipation in the limit of infinite time.

This proof that in the linear response regime the primary dissipation is minimal with respect to variations in the initial distribution function gives a proof (Evans *et al.*, 2015) for T-mixing systems, of Prigogine's principle of minimum entropy production in the linear response regime close to equilibrium. He states: "In the linear regime, the total entropy production in a systems subject to [a] flow of energy and matter, $d_iS/dt = \int \sigma\, dV$, reaches a minimum value at the nonequilibrium stationary state. This is because the unconstrained forces adjust themselves to make their conjugate fluxes go to zero" (Kondepudi and Prigogine, 1998, p. 393). We have already noted that in the linear regime the average dissipation is equal to the so-called entropy production. In our system, there is no net mass flow into or out of the unit cell. In our case all we have to do is to construct a second "force" $F_{e,2}$ that is capable of generating the flux $\dot{g}$. This unconstrained force adjusts itself so that its conjugate flux; namely $\dot{g}$ averages to zero in the steady state. To find this "force" and its equations of motion is a trivial exercise. If the equations of motion take the same form as Eq. (6.12) but with coupling parameters $C_{2,i}, D_{2,i}$ and a "force" $\mathbf{F}_{2,e}$, we see that we merely have to find the coupling parameters such that $\lambda\dot{g} = \mathbf{F}_{2,e} \cdot \sum_{i=1}^N \left[\mathbf{p}_i/m\, D_{2,i} - \mathbf{F}_i C_{2,i}\right]$, which is a trivial exercise.

6.5 Sum Rules for Dissipation in Steady States

Using Eq. (6.15) and the T-mixing property, we have the following relaxation sum rule:

$$\begin{aligned}\lim_{t\to\infty}\langle\dot{g}(t)\rangle_{F_e,\lambda} &= -\int_0^{\infty} ds\langle[\beta_{\text{th}}\mathbf{J}(0)V\cdot\mathbf{F}_e+\lambda\dot{g}(0)]\dot{g}(s)]\rangle_{F_e,\lambda}\\ &= \frac{-\int_0^{\infty} ds\langle[\beta_{\text{th}}\mathbf{J}(0)V\cdot\mathbf{F}_e+\lambda\dot{g}(0)]\dot{g}(s)e^{\lambda g(0)}]\rangle_{F_e,\lambda=0}}{\langle e^{\lambda g(0)}\rangle_{F_e,\lambda=0}}\\ &= 0\end{aligned} \tag{6.18}$$

This is analogous to the corresponding sum rule for the fluxes of nonconserved quantities in systems relaxing to equilibrium – the heat death Eq. (5.36). In the present case, the sum rule is for fluxes of nonconserved quantities relaxing to a steady state. In the heat death case, $F_e = 0$ and the first term on the right-hand side of the first line of Eq. (6.18) is simply absent. So for NESSs in the long time, t, limit, instead of autocorrelation functions of fluxes of nonconserved quantities integrating to zero, they behave as $\lambda\int_0^t ds\ \langle\dot{g}(0)\dot{g}(s)\rangle_{F_e,\lambda}\xrightarrow{\lim(t\to\infty)} -\int_0^t ds\ \langle\beta_{\text{th}}\mathbf{J}(0)V\cdot\mathbf{F}_e\dot{g}(s)\rangle_{F_e,\lambda}$.

The fact that in regard to forming averages of physical phase functions T-mixing systems forget about their initial distributions is completely consistent with our earlier proof that the steady state is physically ergodic. We eventually arrive arbitrarily close to this same domain even if the initial $t = 0$ distribution differs from the equilibrium distribution for the zero-field system. Indeed, if we start a single trajectory at time zero in the long-time limit, the domain traced out by this single trajectory must explore *essentially* the same domain as that generated at some arbitrarily long time from an arbitrary initial distribution of states. This domain will be close to that of the attractor(s) for the system.

Of course, if we examine phase space at extreme resolution (say with a very singular phase function, for example, $\ln[f(\mathbf{\Gamma};t)]$, the deterministic phase space never "forgets" its original initial conditions; these can always be retrieved by applying a time-reversal map to return to the original distribution of states. Therefore, the value of the time average of $\ln[f(\mathbf{\Gamma};t)]$ will depend on the initial point. However, when "observed" by computing averages of low-order physical phase functions for thermophysical properties like pressure, stress, or energy, these very fine structures in phase space cannot be resolved and the knowledge of initial conditions is effectively lost. The measurement of thermophysical properties is the only way we can characterize these macroscopic states. The measurement of the fine-grained phase space density is simply not possible – at least at the resolution required to generate the initial distribution after a long relaxation toward the steady state.

Another way to describe these steady-state strange attractors is that, starting from different initial phase space points, we may approach slightly different steady-state attractors. These different attractors must, in T-mixing systems, be so tightly interwoven that when we measure steady-state averages we cannot observe differences in the long-time averages. This is what is implied by the

T-mixing condition and Eq. (6.18), for example. This happens despite the fact that these attractors are of lower dimension than the ostensible phase space, because there is "room" for them to fill phase space differently, leading to different average values even for low-order phase functions.

At first it may seem paradoxical that, in spite of the fact that the distribution function tends toward a lower dimensional steady-state attractor, the physical averages of properties are independent of the initial distribution of states. It may seem that the dimensional collapse allows plenty of "wriggle room" for physical averages to differ. However, this is countered by two facts. First, in physical T-mixing systems the dimensional reduction is tiny – typically a few parts per Avogadro's number of dimensions! Second, physical properties are dependent only on very low dimensional projections (a dozen or so dimensions) of the full steady-state attractor. These highly projected distributions are smooth.

As long as the steady-state attractor "spans" the ostensible phase space, different initial distributions can yield identical values for physical properties that are only dependent on exceedingly low order projected distributions.

For all temperatures, densities, and external fields, the average long-time dissipation generated from a nonequilibrium distribution is identical to that generated from the equilibrium distribution for the system. This means that the nonlinear transport coefficient $L(F_e)$ defined in terms of the steady state dissipation

$$L(F_e;\lambda) \equiv \lim_{t\to\infty} \frac{\langle \Omega_\lambda(t)\rangle_\lambda}{\beta_{\text{th}} V F_e^2} = \lim_{t\to\infty} \frac{\langle \Omega_{\lambda=0}(t)\rangle_{\lambda=0}}{\beta_{\text{th}} V F_e^2} = L(F_e;\lambda=0) \tag{6.19}$$

is independent of whether the initial distribution was its equilibrium distribution or any deviation from it – so long as the kinetic temperature of the reservoir particles has a fixed value so that the temperature of the underlying equilibrium state is fixed.

Equation (6.19) accords with our knowledge of the thermophysical properties of fluids, and so on. For example, the viscosity of argon is history-independent. It only depends of the temperature, density, and strain rate. The initial preparation of the system is irrelevant to the viscous properties of the system in the steady state inside the viscometer. In T-mixing systems, the nonlinear and linear transport coefficients are in fact state functions.

6.6
Positivity of Nonlinear Transport Coefficients

In Chapter 4, we gave a derivation of the dissipation theorem for an exceedingly general set of time-reversible equations of motion and for quite general initial distributions $f(\mathbf{\Gamma};0)$. Considering a nonequilibrium system, if one substitutes Ω for B in Eq. (6.8) and then combines the resulting equation with the strong form of the second law inequality, one knows that time integrals of the ensemble averages of the dissipation must be positive: $\int_0^t ds\ \langle\Omega(s)\rangle > 0,\ \ \forall t$. Since at long times for T-mixing systems the average dissipation is time-independent, one can only conclude that in NESSs the ensemble average dissipation must be positive. If this were not the case, the second law inequality would be violated for sufficiently large times. Therefore, the dissipation in *driven* T-mixing systems

$$\lim_{t\to\infty} \langle \Omega(t)\rangle_{F_e,f(\boldsymbol{\Gamma};0)} = \int_0^\infty ds \langle \Omega(0)\Omega(s)\rangle_{F_e,f(\boldsymbol{\Gamma};0)} = const > 0, \quad \forall F_e,\ f(\boldsymbol{\Gamma};0) \tag{6.20}$$

So, for driven systems not only does the dissipation autocorrelation function start with a positive value ($\langle \Omega(0)^2\rangle_{F_e,f(\boldsymbol{\Gamma};0)} > 0,\ \forall F_e, f(\boldsymbol{\Gamma};0)$), but for all normalizable initial distributions and for any well-defined dynamics with an arbitrarily strong external field (if any), any negative tails in the ensemble-averaged dissipation function must disappear before the system enters the necessarily positive dissipation of the final steady state.

If we consider a driven isokinetic system, we observe from Eq. (6.8) that

$$\begin{aligned} -\lim_{t\to\infty}\langle J(t)\rangle_{F_e,c} &= -\langle J(0)\rangle_c + \beta_{\rm th} V \int_0^\infty ds \langle J(0)J(s)\rangle_{F_e,c} F_e \\ &= \beta_{\rm th} V \int_0^\infty ds \langle J(0)J(s)\rangle_{F_e,c} F_e \\ &\equiv L(F_e)F_e > 0, \quad \forall F_e \end{aligned} \tag{6.21}$$

We have assumed the dissipative flux and force are scalars, and we have used the fact that $\beta_{\rm th} \equiv 1/k_{\rm B}T_{\rm th}$, where $T_{\rm th}$ is the equilibrium thermodynamic temperature that the system will relax to if the driving force is set to zero and the system is allowed to relax to equilibrium. We have also used the fact that for driven systems $\langle J(0)\rangle_c = 0$.

The T-mixing property guarantees that the $t \to \infty$ limit is finite and therefore so too is the nonlinear transport coefficient at the specified value of the driving field $L(F_e)$. The T-mixing condition further guarantees that the NESS is physically ergodic over the specified phase space domain.

The second law inequality means that the conventionally defined average dissipative flux will be negative when the dissipative field is positive. If we consider planar Couette flow as an example, the following mapping applies: $F_e \to \partial u_x/\partial y \equiv \dot{\gamma}$; $J \to P_{xy}$; $L(F_e) \to \eta(\dot{\gamma})$, where the variables are, in turn, the strain rate $\dot{\gamma}$; the xy element of the pressure tensor P_{xy}, and, lastly, the nonlinear strain-rate-dependent shear viscosity $\eta(\dot{\gamma})$ defined in the *nonlinear constitutive relation* for shear viscosity. Equation (6.21) gives a nonlinear constitutive relation between the dissipative flux and the dissipative field for a finite driven thermostatted system that is T-mixing.

The second law inequality guarantees that the nonlinear transport coefficients $L(F_e), \eta(\gamma)$ appearing in the nonlinear constitutive relation for finite-sized periodic, T-mixing systems are finite and must be positive:

$$\infty > L(F_e) > 0, \quad \forall F_e \tag{6.22}$$

If we look again at Eq. (6.21), we see that the second law inequality implies that $\int_0^t ds \langle J(s)\rangle_{F_e,f_c(\boldsymbol{\Gamma},0)} < 0,\ \forall t$. However, the approach to the steady state may not be monotonic. The ensemble-averaged instantaneous dissipative flux may be positive at intermediate times. In fact, in the nonlinear regime this is a common situation.

6.7 Linear Constitutive Relations for T-Mixing Canonical Systems

Each of these results also includes the linear response regime as a special case in the limit of weak fields.

$$
\begin{aligned}
&\lim_{F_e\to 0}\lim_{t\to\infty}\frac{-\partial\langle J(t)\rangle_{F_e,c}}{\partial F_e}\\
&= \lim_{F_e\to 0}\beta_{\rm th}V\int_0^\infty ds\left\langle J(0)\frac{\partial J(s)}{\partial F_e}\right\rangle_{F_e,c}F_e + \beta_{\rm th}V\int_0^\infty ds\langle J(0)J(s)\rangle_{F_e=0,c}\\
&= \lim_{F_e\to 0}\beta_{\rm th}V\int_0^\infty ds\langle J(0)J'(s)\rangle_{F_e,c}F_e + \beta_{\rm th}V\int_0^\infty ds\langle J(0)J(s)\rangle_{F_e=0,c}\\
&= \beta_{\rm th}V\int_0^\infty ds\langle J(0)J(s)\rangle_{F_e=0,c}\\
&\equiv L(F_e=0) > 0
\end{aligned}
\tag{6.23}
$$

In Eq. (6.23), there are two places where the field dependence is manifest. One is in the explicit factor F_e. The second place is in the implicit time dependence of $J(s)$. In going from the first to the second line of Eq. (6.23), we expect that since $J(s)$ is a smooth function of time, phase $\boldsymbol{\Gamma}$, and F_e, $J'(\boldsymbol{\Gamma}(s)) \equiv \partial J(\boldsymbol{\Gamma}(s))/\partial F_e$ will also be smooth. We assume the equations of motion do not contain singularities. We do not cover the case of hard particles or even systems with a piecewise continuous potential.

We note that $d\langle J(t)\rangle/dt = \beta_{\rm th}VF_e\langle J(0)J(t)\rangle_{F_e,c}$, so at short times $\lim_{t\to 0}\langle J(t)\rangle_{F_e,c} = \beta_{\rm th}V\langle J(0)J(0)\rangle_{0,c}F_e t$, which is linear in the field, and means that it takes time for the nonlinearities to "grow" into the response.

The proof of Eq. (6.23) gives a proof that finite systems that are thermostatted and driven satisfy AI$\boldsymbol{\Gamma}$ with smooth intermolecular forces, are T-mixing, and have finite linear constitutive relations in the weak field limit. Further, the transport coefficient appearing in this linear constitutive relation is positive. For electrical conductivity, we therefore have, subject to the conditions above, a proof of Ohm's "Law" in the limit of weak fields, or using the SLLOD equations for shear flow, a proof of Newton's constitutive relation for weak shear flow – at least as they apply to finite periodic systems. There is nothing in the proof given above to prevent the possible divergence of the extrapolated linear transport coefficient as the system size, in each periodic cell, is increased.

6.8 Gaussian Statistics for T-Mixing NESS

In 2000, we showed (Searles and Evans, 2000) that by combining the asymptotic steady-state Evans–Searles transient fluctuation theorem (ESFT) (proved in the next section) with the central limit theorem, you could prove Green–Kubo relations in driven systems, for transport coefficients in the weak-field limit.

That derivation required a careful double limit ($t \to \infty, F_e \to 0$) that could not be extended to higher field strengths. It also required the *assumption* that for long averaging times the time-averaged dissipative flux satisfies the conditions for the central limit theorem to be valid. The dissipation theorem obviates this discussion and shows how both the nonlinear and the linear response can be obtained directly and exactly in terms of integrals of time correlation functions involving the dissipation function directly.

However, if the system is ΩT-mixing (or mixing), then for sufficiently long averaging times the time-averaged dissipative flux *must* satisfy the central limit theorem with Gaussian statistics close to the mean of the distribution. As we will see in the next section, this fact is essential in order to prove the observability of the asymptotic steady-state fluctuation relation.

For driven systems, the dissipation function is quite simply related to the dissipative flux and the dissipative force. The Green–Kubo equilibrium time correlation function involves fluctuations in the dissipative flux. This flux is *not* identically zero at any instant in time for a system at equilibrium, whereas the dissipation function is. For driven systems satisfying AI$\mathbf{\Gamma}$, the dissipative force, and not the flux, is zero at equilibrium at all times.

6.9
The Nonequilibrium Steady-State Fluctuation Relation

We now consider fluctuation relations for the dissipation in a NESS (Searles, Rondoni, and Evans, 2007) – or at least as we approach NESSs. We have already seen (Section 6.4) that, if the initial distribution is not the equilibrium distribution for the zero-field dynamics, the influence of the deviation function disappears in time Eq. (6.4.5). This means, by definition, that any steady-state fluctuation relation can *only* refer to the primary dissipation function for the system.

From Section 3.6, we may approach the steady state by asking what is the probability that the covariant dissipation integrated for a time τ, but starting not at time zero but rather at time t, equals a value A compared to $-A$. As t becomes ever larger, the time-integrated dissipation approaches that of a true NESS. So using Eqs. (3.26) and (3.29) we can write down the following exact Evans–Searles transient fluctuation relation for the time-averaged dissipation function defined by the system dynamics and the time t distribution function $\overline{\Omega}_{t,t+\tau}(S^t\mathbf{\Gamma};t) = \frac{1}{\tau}\Omega_{t,t+\tau}(S^t\mathbf{\Gamma};t)$. To simplify the notation, we let $\overline{\Omega}_{u,u+\tau}(S^u\mathbf{\Gamma};t) \equiv \overline{\Omega}_{u,u+\tau}(t)$:

$$\frac{1}{\tau}\ln\left[\frac{p\left(\overline{\Omega}_{t,t+\tau}(t)=A\right)}{p(\overline{\Omega}_{t,t+\tau}(t)=-A)}\right] = \frac{1}{\tau}\ln\left[\frac{p(\overline{\Omega}_{0,2t+\tau}(0)=A\tau/(2t+\tau)}{p(\overline{\Omega}_{0,2t+\tau}(0)=-A\tau/(2t+\tau))}\right]$$
$$= A\tau/(2t+\tau) \tag{6.24}$$

As discussed in Section 3.6, the covariant steady-state fluctuation relation reduces to the familiar transient fluctuation relation over a symmetrically extended time range. On the right-hand side of Eq. (6.24), the dissipation function is defined with respect to the time zero distribution, $\overline{\Omega}_{0,2t+\tau}(\mathbf{\Gamma};0)$.

We can determine the steady-state distribution by taking the long-time limit $p(\overline{\Omega}_{\mathrm{ss},\tau}(\mathrm{ss}) = A) \equiv \lim_{t\to\infty} p(\overline{\Omega}_{t,t+\tau}(t) = A)$, where $\overline{\Omega}_{ss,\tau}(ss)$ is the dissipation function defined with respect to the steady state and averaged over a period τ, starting in the steady state. Therefore

$$\frac{1}{\tau}\ln\left[\frac{p\left(\overline{\Omega}_{\mathrm{ss},\tau}(\mathrm{ss}) = A\right)}{p(\overline{\Omega}_{\mathrm{ss},\tau}(\mathrm{ss}) = -A)}\right] \equiv \lim_{t\to\infty}\frac{1}{\tau}\ln\left[\frac{p\left(\overline{\Omega}_{t,t+\tau}(t) = A\right)}{p(\overline{\Omega}_{t,t+\tau}(t) = -A)}\right]$$
$$= A \tag{6.25}$$

One can also ask what conditions are required for a steady-state fluctuation relation to be valid when the dissipation function is defined with respect to the zero time distribution rather than the evolved or steady-state distribution.

First, we discuss the case where there is no serial correlation in the time series data for $\Omega(S^t\mathbf{\Gamma};0)$. Using Eq. (3.25), we can write

$$\begin{aligned}
\lim_{t\to\infty}\frac{1}{\tau}\ln\frac{p(\overline{\Omega}_{t,t+\tau}(0) = A)}{p(\overline{\Omega}_{t,t+\tau}(0) = -A)} &= \lim_{t\to\infty}\frac{1}{\tau}\ln\left[\left\langle\exp(-\Omega_{0,2t+\tau}(0)\right\rangle^{-1}_{\Omega_{t,t+\tau}(0)=A\tau}\right] \\
&= \lim_{t\to\infty}\frac{1}{\tau}\ln\left[\left\langle\exp\left(-\Omega_{0,t}(0) - \Omega_{t,t+\tau}(0)\right.\right.\right. \\
&\quad \left.\left.\left.-\Omega_{t+\tau,2t+\tau}(0)\right)\right\rangle^{-1}_{\Omega_{t,t+\tau}(0)=A\tau}\right] \\
&= A + \lim_{t\to\infty}\frac{1}{\tau}\ln\left[\left\langle\exp\left(-\Omega_{0,t}(0)\right.\right.\right. \\
&\quad \left.\left.\left.-\Omega_{t+\tau,2t+\tau}(0)\right)\right\rangle^{-1}\right] \\
&= A + \lim_{t\to\infty}\frac{1}{\tau}\ln\left[\left\langle\exp\left(-\Omega_{0,t}(0)\right)\right\rangle^{-1}\right. \\
&\quad \left.\left\langle\exp\left(-\Omega_{t+\tau,2t+\tau}(0)\right)\right\rangle^{-1}\right] \\
&= \lim_{t\to\infty}\left[A + \frac{1}{\tau}\ln\left[\left\langle\exp\left(-\Omega_{t+\tau,2t+\tau}(0)\right)\right\rangle^{-1}\right]\right]
\end{aligned} \tag{6.26}$$

where we have exploited the lack of serial correlation in the data and employed the nonequilibrium partition identity Eq. (3.12). If the time series data for $\Omega(t)$ has no serial correlation, we can see that the condition for the ensemble average has no influence on the time integrals inside the ensemble average on the third line of Eq. (6.26).

The second term in the last line of Eq. (6.26) can be evaluated using the relations

$$\begin{aligned}\left\langle \exp\left(-\Omega_{0,a+b}(0)\right)\right\rangle &= \left\langle \exp\left(-\Omega_{0,a}(0)\right)\right\rangle \left\langle \exp\left(-\Omega_{a,a+b}(0)\right)\right\rangle = 1\\ &= 1\left\langle \exp\left(-\Omega_{a,a+b}(0)\right)\right\rangle\\ &\Rightarrow \left\langle \exp\left(-\Omega_{a,a+b}(0)\right)\right\rangle = 1\end{aligned} \tag{6.27}$$

Substituting this into Eq. (6.26) gives

$$\begin{aligned}\lim_{t\to\infty}\frac{1}{\tau}\ln\frac{p(\overline{\Omega}_{t,t+\tau}(0)=A)}{p(\overline{\Omega}_{t,t+\tau}(0)=-A)} &\equiv \frac{1}{\tau}\ln\frac{p(\overline{\Omega}_{\mathrm{ss},\tau}(0)=A)}{p(\overline{\Omega}_{\mathrm{ss},\tau}(0)=-A)}\\ &= A\end{aligned} \tag{6.28}$$

where the subscript "ss" in $\overline{\Omega}_{\mathrm{ss},\tau}(0)$ denotes that the time averages start in the steady state, and the zero in the argument denotes the fact that the dissipation is defined with respect to the zero time (equilibrium) distribution.

Now, of course, in any real dynamical system there must be serial correlation in the time series data, so Eq. (6.28) cannot be exact for real dynamical systems. The covariant properties of the dissipation function mean that extreme care is needed when attempting to handle these equations for NESSs. When the serial correlation is allowed for, Eq. (6.28) becomes an asymptotic result that is valid only in the limit $\tau/\tau_M \to \infty$, where τ_M is the Maxwell time describing the correlation time of the dissipation function:

$$\begin{aligned}\lim_{\tau/\tau_M\to\infty}\lim_{t\to\infty}\frac{1}{\tau}\ln\left[\frac{p\left(\overline{\Omega}_{t,t+\tau}(0)=A\right)}{p(\overline{\Omega}_{t,t+\tau}(0)=-A)}\right] &= \lim_{\tau/\tau_M\to\infty}\frac{1}{\tau}\ln\left[\frac{p\left(\overline{\Omega}_{\mathrm{ss},\tau}(0)=A\right)}{p(\overline{\Omega}_{\mathrm{ss},\tau}(0)=-A)}\right]\\ &= A\end{aligned} \tag{6.29}$$

where the subscript "ss" denotes that the integral over τ should only be done when, to your desired level of accuracy, the system has relaxed to its unique NESS and the dissipation function is the primary dissipation function for the dynamics. We will now derive this result.

For sufficiently large t, that is, several Maxwell times τ_M, we can approach a steady state. Observing

$$\begin{aligned}\overline{\Omega}_{t,t+\tau}(0) &= \frac{1}{\tau}\int_t^{t+\tau} ds\ \Omega(S^s\boldsymbol{\Gamma};0)\\ &= \frac{1}{\tau}\int_0^{\tau} ds\ \Omega(S^s\boldsymbol{\Gamma};0) + \frac{1}{\tau}\int_0^{t} ds(\Omega(S^{\tau+s}\boldsymbol{\Gamma};0) - \Omega(S^s\boldsymbol{\Gamma};0))\end{aligned} \tag{6.30}$$

and noting that last integral on the second line does not grow/shrink with time after the first few Maxwell times, we can write $\overline{\Omega}_{t,t+\tau} = \overline{\Omega}_{0,\tau} + O(\tau_M/\tau)$ and

$\lim_{\tau/\tau_M\to\infty}\overline{\Omega}_{t,t+\tau} = \overline{\Omega}_{0,\tau}$. So $\lim_{\tau/\tau_M\to\infty} p(\overline{\Omega}_{t,t+\tau}(0) = A) = \lim_{\tau/\tau_M\to\infty} p(\overline{\Omega}_{0,\tau}(0) = A + O(\tau_M/\tau))()$, $\forall t$ Therefore

$$\begin{aligned}
\lim_{\tau/\tau_M\to\infty} \frac{1}{\tau} \ln \frac{p(\overline{\Omega}_{ss,\tau}(0) = A)}{p(\overline{\Omega}_{ss,\tau}(0) = -A)} &= \lim_{\tau/\tau_M\to\infty} \lim_{t\to\infty} \frac{1}{\tau} \ln \frac{p(\overline{\Omega}_{t,t+\tau}(0) = A)}{p(\overline{\Omega}_{t,t+\tau}(0) = -A)} \\
&= \lim_{\tau/\tau_M\to\infty} \frac{1}{\tau} \ln \frac{p(\overline{\Omega}_{0,\tau}(0) = A + O(\tau_M/\tau))}{p(\overline{\Omega}_{0,\tau}(0) = -A + O(\tau_M/\tau))}, \forall t \\
&= \lim_{\tau/\tau_M\to\infty} [A + O(\tau_M/\tau)] \\
&= A
\end{aligned} \tag{6.31}$$

The second line transforms the steady-state probability ratio into a ratio of transient averages of dissipation. We note that, in order for the proof to be valid, the standard deviation of $\overline{\Omega}_{t,t+\tau}$ should shrink more slowly than $1/\tau$. Otherwise, the corrections $O(\tau_M/\tau)$ will not become negligible before observation of fluctuations becomes impossible.

In Eq. (6.31), the dissipation function is defined by the dynamics and the $t = 0$ (equilibrium) distribution function. We only *integrate* the resulting function over the restricted range $t, t + \tau$, where t is sufficiently long that we are in the steady state. We have already seen that for T-mixing systems the system eventually forgets about the initial distribution. By definition, this process must be complete before a NESS can be created.

As we have noted in Section 6.8, for T-mixing steady states the distribution of average values of dissipation will become Gaussian about the mean. Since the average dissipation is positive, the negative value that is most difficult to observe is -1 times the mean value of the dissipation $\overline{A}$. The mean will remain constant with increase in averaging time t, but the standard deviation will decrease. Using Gaussian statistics we see that $\overline{A} = |-\overline{A}| \sim \sigma_A O(\tau^{-1/2})$, where σ_A is the standard deviation of the distribution of A. Taking more and more samples enables us to observe fluctuations further and further from the mean value for Ω – which is positive.

Definition

So the asymptotic steady-state fluctuation relation (6.31) is *observable* because the error term $(= O(\tau^{-1}))$ vanishes faster than the (negative) fluctuations themselves $(= O(\tau^{-1/2}))$. This means that the fluctuations satisfy the asymptotic steady-state fluctuation relation at long averaging times *before* the magnitude of those fluctuations decays to zero.

For T-mixing systems, the steady state is physically ergodic and independent of the initial distribution of states. If we take the initial distribution to arbitrarily close to a delta function at almost any point in phase space, our asymptotic steady-state fluctuation relation applies to late-time averages along a single phase space trajectory. In this case, Eq. (6.31) is an asymptotic result for an individual dynamical system over arbitrarily long times.

6.10 Gallavotti–Cohen Steady-State Fluctuation Relation

An alternative steady-state fluctuation relation to Eq. (6.31) has been proposed by Gallavotti and Cohen (1995a,b) and Bonetto *et al.* (2006). The Gallavotti–Cohen fluctuation relation has been *proven* for Anosov and the so-called Axiom A systems (Gallavotti, 1995), but the resulting relationship was anticipated to apply to a wider range of systems.

The Gallavotti–Cohen fluctuation relation can be written as

$$\lim_{\tau\to\infty} \frac{1}{\tau} \ln \left[\frac{p[\overline{\Lambda}_\tau = B]}{p[\overline{\Lambda}_\tau = -B]} \right] = -B \quad \text{for } |B| \le B^* \tag{6.32}$$

where $\Lambda \equiv {}^{\partial}\!/\!_{\partial\Gamma} \cdot \dot{\Gamma}$ is the phase space expansion rate and B^* is some bound (generally unknown) (Evans, Searles, and Rondoni, 2005). Equation (6.32) refers to results observed along a single, exceedingly long phase space trajectory.

Equation (6.32) has, as we have said, been proven for Anosov systems, but Gallavotti and Cohen proposed that the equation may be valid for sufficiently chaotic non-Anosov systems. This proposal is termed the *chaotic hypothesis*. At the present time there is no test, independent of the Gallavotti–Cohen fluctuation relation, that predicts whether the chaotic hypothesis will apply to any given non-Anosov system. Presumably, a precondition for the chaotic hypothesis to hold is that the dynamical system does in fact relax to a steady state because Eq. (6.32) would make no sense for non-steady-state systems (e.g., adiabatic systems that heat up without bound).

Gallavotti has also proposed a possible modification to Eq. (6.32) for systems with an unbalanced number of positive and negative Lyapunov exponents in non-Anosov systems (Bonetto, Gallavotti, and Garrido, 1997). However, numerical tests seem to show no evidence of a discontinuity in Eq. (6.32) when the number of positive and negative exponents change (e.g., by increasing the dissipative field). Such a change would be necessarily discontinuous (Williams, Searles, and Evans, 2006).

For isoenergetic systems, $\Lambda(t) = -\Omega(t),\ \forall t$, and therefore the relations (6.29) and (6.32) become identical for ergodic, isoenergetic steady states, implying for this circumstance that $B^* = \infty$.

Application of the Gallavotti–Cohen fluctuation relation to systems that are not isoenergetic has recently been discussed (Evans, Searles, and Rondoni, 2005; Bonetto *et al.*, 2006), and it has found that there are serious limitations to its practical utility. For instance, for systems driven by a dissipative field F_e and satisfying AIΓ, the bound in Eq. (6.32) goes to zero as equilibrium is approached: $B^* = O(F_e^2) \to 0$ as $F_e \to 0$. This means that the range of applicability of the Gallavotti-Cohen fluctuation relation shrinks to zero as equilibrium is approached. In fact, it is easy to see why this must be the case. At equilibrium, the Gallavotti-Cohen fluctuation relation for thermostatted systems would predict an asymmetry in the probability of time-averaged values of the phase space expansion factor. This is obviously not possible! By contrast, at equilibrium the Evans–Searles fluctuation relations simply state that fluctuations in the time-integrated dissipation is symmetric about zero – see Section 6.9.

Perhaps even more difficult is the fact that for thermostatted systems the time required for convergence of the Gallavotti–Cohen fluctuation relation diverges to infinity as $O(F_e^{-2})$ (i.e., the asymptotic limit in Eq. (6.32) should be written as $\tau/F_e^2 \to \infty$) (Evans, Searles, and Rondoni, 2005).

Since much of the interest in fluctuation relations arises from the fact that they are exact arbitrarily far from equilibrium, the bound on the range of fluctuations means that the Gallavotti–Cohen fluctuation relation is of limited use in large deviation theory. On the other hand, close to equilibrium the shrinking bounds on the range of the argument and the divergence of the convergence time also lead to problems.

One can easily see why this divergence of convergence times occurs close to equilibrium. The phase space expansion factor for thermostatted systems close to equilibrium contains a sum of two terms. One is the dissipation function (times −1) but the other component is just (to leading order, close to equilibrium) the equilibrium fluctuations in the phase space expansion factor. The equilibrium fluctuations become independent of the external field close to equilibrium, and they are of course symmetric about zero and therefore *cannot* satisfy any fluctuation relation. In the long-time limit in steady states of thermostatted systems $\lim_{t\to\infty}(\overline{\Lambda}_t + \overline{\Omega}_t) = 0$, but as the field becomes ever smaller, the relative magnitude of the symmetric equilibrium fluctuations becomes ever larger, swamping the dissipation. Thus as the field becomes smaller, it takes longer and longer for the average $\overline{\Lambda}_t$ to become equal to $-\overline{\Omega}_t$. We will discuss an example of the convergence difficulties for the Gallavotti–Cohen fluctuation relation in Section 7.3 – especially contrasting Eqs. (7.27)–(7.31).

6.11 Summary

One often sees references in the literature to the supposition that in NESSs the "entropy production" (i.e., average dissipation) is a maximum (or sometimes a minimum!) subject to the known constraints. The fact that when a dissipative field is suddenly applied to an equilibrium system the dissipation increases from zero means that in a steady state the dissipation can hardly be an absolute minimum. The fact that the dissipation very frequently overshoots its steady state value means that, in general, the steady-state dissipation cannot be a maximum either. In this chapter, we have shown that in the linear response regime the primary dissipation is minimal with respect to all possible variations of the initial distribution away from the natural equilibrium distribution.

There is a way of rederiving the dissipation theorem for driven systems as an extremum principle (Evans, 1985), but the final result is identical to the dissipation theorem and it involves an *infinite* set of constraints. The choice of which constraints should be used in these derivations is best made *after* you already know the correct answer, because *a priori* there seems to be no objective criteria for selecting these constraints.

One of the interesting things our work has revealed is that in T-mixing systems the NESS is physically ergodic and independent of the initial distribution. This independence with respect to the initial distribution means that there is only one Evans–Searles steady-state fluctuation relation for a given dynamical system.

The positivity of transport coefficients is a direct result of the fact that the time integral of the ensemble-averaged dissipation is positive. It also means that, on average, work is converted into heat rather than the reverse. For a driven T-mixing system that satisfies AIΓ and is isokinetic, we have

$$\begin{aligned}\beta_{\text{th}}\dot{H}_0(t) &= -\beta_{\text{th}} VJ(t)F_e - 2\beta_{\text{th}}K_{\text{th}}\alpha(t) \\ &= -\beta_{\text{th}} VJ(t)F_e - (3N-4)\alpha(t)\end{aligned} \tag{6.33}$$

If we take long-time averages for a steady state

$$\begin{aligned}\lim_{t\to\infty} \beta_{\text{th}}\overline{\dot{H}_{0,t}} &= \lim_{t\to\infty}\left[-\beta_{\text{th}}\overline{J}_t VF_e - (3N-4)\,\overline{\alpha}_t\right] \\ &= \beta_{\text{th}} VL(F_e)F_e^2 - (3N-4)\lim_{t\to\infty}\overline{\alpha}_t \\ &= 0\end{aligned} \tag{6.34}$$

where $\overline{\cdots}_t$ denotes a time average of duration t. We note that it is the stationary property of the ΩT-mixing steady-state which implies that the long-time average rate of change of the energy goes to zero. Since $\beta_{\text{th}}, L(F_e), V, F_e^2$ are each strictly positive, so too must be the long-time average of the thermostat multiplier. This means that when averaged over long times in a NESS, heat must, on average, be removed from the system by the thermostat. Thus the work performed on the system by the dissipative field $-\overline{J}VF_e$ is, on average, positive and by Eq. (6.34) this work is *dissipated* into the form of heat and then removed from the system by the (physically remote) thermostat. This gives a mathematical proof of one of the postulated statements of the second law of thermodynamics given in William Thomson's, 1852 paper, "On the universal Tendency in Nature to the Dissipation of Mechanical Energy" (Thomson, 1852): "Although mechanical energy is indestructible, there is a universal tendency to its dissipation, which produces throughout the system a gradual augmentation and diffusion of heat, cessation of motion, and exhaustion of the potential energy of the material Universe."

References

Alder, B.J. and Wainwright, T.E. (1970) Decay of the velocity autocorrelation function. *Phys. Rev. A*, **1**, 18–21.

Allen, M.P. and Tildesley, D.J. (1987) *Computer Simulation of Liquids*, Clarendon Press, Oxford.

Bonetto, F., Gallavotti, G., and Garrido, P.L. (1997) Chaotic principle: an experimental test. *Physica D*, **105**, 226–252.

Bonetto, F., Gallavotti, G., Giuliani, A., and Zamponi, F. (2006) Chaotic hypothesis, fluctuation theorem and singularities. *J. Stat. Phys.*, **123**, 39–54.

Evans, D.J. (1985) Response theory as a free-energy extremum. *Phys. Rev. A*, **32**, 2923–2925.

Evans, D.J., Cohen, E.G.D., Searles, D.J., and Bonetto, F. (2000) Note on the Kaplan-Yorke dimension and linear transport coefficients. *J. Stat. Phys.*, **101**, 17–34.

Evans, D.J. and Morriss, G.P. (1990) *Statistical Mechanics of Nonequilibrium Liquids*, Academic Press, London.

Evans, D.J. and Sarman, S. (1993) Equivalence of thermostatted nonlinear responses. *Phys. Rev. E*, **48**, 65–70.

Evans, D.J., Searles, D.J., and Rondoni, L. (2005) Application of the Gallavotti-Cohen fluctuation relation to thermostated steady states near equilibrium. *Phys. Rev. E*, **71**, 056120.

Evans, D.J., Williams, S.R., Searles, D.J., and Rondoni, L. (2015) On the relaxation to nonequilibrium steady states. submitted, *J. Stat. Phys.*, arXiv:1602.05808 [cond-mat.stat-mech].

Gallavotti, G. (1995) Reversible Anosov maps and large deviations. *Math. Phys. Electron. J.*, **1**, 1–12.

Gallavotti, G. and Cohen, E.G.D. (1995a) Dynamical ensembles in nonequilibrium statistical mechanics. *Phys. Rev. Lett.*, **74**, 2694–2697.

Gallavotti, G. and Cohen, E.G.D. (1995b) Dynamical ensembles in stationary states. *J. Stat. Phys.*, **80**, 931–970.

Huang, K. (1963) *Statistical Mechanics*, John Wiley & Sons, Inc., New York.

Kondepudi, D. and Prigogine, I. (1998) *Modern Thermodynamics, from Heat Engines to Dissipative Structures*, John Wiley & Sons, Ltd., Chichester.

Kubo, R. (1966) The fluctuation-dissipation theorem. *Rep. Prog. Phys.*, **29**, 255.

McQuarrie, D.A. (1976) *Statistical Mechanics*, Harper and Row, New York.

Searles, D.J. and Evans, D.J. (2000) The fluctuation theorem and Green-Kubo relations. *J. Chem. Phys.*, **112**, 9727–9735.

Searles, D.J., Evans, D.J., and Isbister, D.J. (1997) The number dependence of the maximum Lyapunov exponent. *Physica A*, **240**, 96–104.

Searles, D.J., Rondoni, L., and Evans, D.J. (2007) The steady state fluctuation relation for the dissipation function. *J. Stat. Phys.*, **128**, 1337–1363.

Thomson, W. (1852) On a universal tendency in nature to the dissipation of mechanical energy. *Proc. R. Soc. Edinburgh*, **3**, 139–142.

Thomson, W. (1874) Kinetic theory of the dissipation of energy. *Nature*, **9**, 441.

Williams, S.R., Searles, D.J., and Evans, D.J. (2006) Numerical study of the steady state fluctuation relations far from equilibrium. *J. Chem. Phys.*, **124**, 194102-1.

Williams, S.R., Searles, D.J., and Evans, D.J. (2014) On the relationship between dissipation and the rate of spontaneous entropy production from linear irreversible thermodynamics. *Mol. Simul.*, **40**, 208–217.

7
Applications of the Fluctuation, Dissipation, and Relaxation Theorems

M. Van der Waals seems, therefore, to be somewhat hasty in assuming that the temperature of a substance is in every case measured by the energy of agitation of its individual molecules.

(Maxwell, 1874)

7.1
Introduction

In this chapter we will work through a number of applications of the various theorems we have proven earlier in this book. These applications illustrate the power of the formalism we have constructed. The first application will be a derivation of the zeroth "Law" of thermodynamics (Evans, Williams, and Rondoni, 2012). It is ironic that the first statement of the zeroth "Law" was by Maxwell. The quote above suggests that he was quite confused regarding the velocity distributions of equilibrium systems at different densities. However, Maxwell's concept of the zeroth "Law" was entirely macroscopic and had nothing to do with microscopic distributions.

Later we treat heat flow (Searles and Evans, 2001) and temperature quenches (Schmidt and Evans, 1994) showing how these transient thermal transport processes can be expressed in terms of dissipative fluxes and fields. Heat flow is interesting because it is a boundary-driven *thermal*, rather than a mechanical, transport process. There is no mechanical dissipative field appearing in the equations of motion – as is the case for electrical conductivity or the SLLOD equations for shear flow. Similarly, a temperature quench has no mechanically dissipative field. We also cover the relaxation of inhomogeneities in atomic "color." If the color of a particle is ignored, in the weak field regime, the system is always in equilibrium (Evans, Searles, and Mittag, 2001). The system is in a nonequilibrium state only if we can recognize the "color" of its particles. The equations of motion for this system are field-free Hamiltonian dynamics. No energy is exchanged between the system of interest and its surroundings. Nevertheless, we can calculate the dissipation function and evaluate fluctuation relations!

Fundamentals of Classical Statistical Thermodynamics: Dissipation, Relaxation and Fluctuation Theorems, First Edition. Denis J. Evans, Debra J. Searles, and Stephen R. Williams.

Later we give a very brief discussion of instantaneous fluctuation relations (Petersen, Evans, and Williams, 2013) and, finally, a summary of some of the fundamental properties of the dissipation function and contrast these with the phase space expansion factor.

7.2 Proof of the Zeroth "Law" of Thermodynamics

In this section, we turn our attention to the proof of the zeroth "Law" of thermodynamics (Evans, Williams, and Rondoni, 2012). What is now known as the *zeroth law of thermodynamics* was first stated by Maxwell (1872). Among the numerous equivalent statements, Maxwell said: "Bodies whose temperatures are equal to that of the same body have themselves equal temperatures." In Chapter 5, we gave a number of proofs of the relaxation of classical particulate systems to thermal equilibrium. The equilibrium states to which our systems relax are all spatially isothermal. None of the equilibrium distributions refers to the absolute positions of particles. They are only a function of their relative positions through the interatomic potential energy function. Implicit in these equilibrium relaxation theorems is a proof of the zeroth "Law" of thermodynamics.

In this section we give an explicit mathematical proof of the zeroth "Law" for T-mixing deterministic particulate systems obeying autonomous Hamiltonian dynamics. No external fields are applied to the system. We should add that, as in most discussions in physics, we consider only inertial coordinate systems since we do not wish to include Coriolis forces, and so on.

The derivation also leads to an understanding of how heat flows from hot to cold bodies and how the transport coefficient characterizing this flow is positive and finite when the system is T-mixing. This heat flow gradually equalizes the temperature across the entire system, and heat eventually ceases to flow. This is the mechanism by which the zeroth "Law" behavior of equilibrium systems is achieved.

Consider an ensemble of N-particle systems obeying Newton's or Hamilton's equations of motion. We do not assume that each particle is identical. The particles could differ in masses and interatomic potentials. If the system of particles is isolated, the total energy, linear momentum, and angular momentum are constants of the motion. In our thought experiment, we could imagine that the system is composed of two solid three-dimensional boxes so that the left half and the right half of the system are in thermal contact but there is no mass flow between the two sides. These two boxes represent the "bodies" mentioned in Maxwell's statement of the zeroth "Law." The two boxes (bodies) contain particles that may be solid, liquid, or gas.

In order to prove the zeroth "Law" of thermodynamics, consider a system with different temperatures in its left and right sides. We let $\Delta\beta$ denote the difference in the reciprocal absolute temperatures of the two bodies left L and right R, divided by Boltzmann's constant. The absolute temperature of each body, T_L, T_R, is not known. Only the reciprocal difference $\Delta\beta \equiv {}^1\!/_{k_B T_R} - {}^1\!/_{k_B T_L}$ is known. The calculation of the absolute temperature of each body requires knowledge of the equation of state for the system. Thus we consider an initial distribution of the form

$$f(\mathbf{\Gamma};0) = \frac{\exp[\Delta\beta H_L(\mathbf{\Gamma})]\delta(H_0(\mathbf{\Gamma})-E)\delta(\mathbf{P})\delta(\mathbf{L})}{\int d\mathbf{\Gamma}\exp[\Delta\beta H_L(\mathbf{\Gamma})]\delta(H_0(\mathbf{\Gamma})-E)\delta(\mathbf{P})\delta(\mathbf{L})}, \quad \Delta\beta \neq 0$$
$$= \frac{\exp[-\Delta\beta H_R(\mathbf{\Gamma})]\delta(H_0(\mathbf{\Gamma})-E)\delta(\mathbf{P})\delta(\mathbf{L})}{\int d\mathbf{\Gamma}\exp[-\Delta\beta H_R(\mathbf{\Gamma})]\delta(H_0(\mathbf{\Gamma})-E)\delta(\mathbf{P})\delta(\mathbf{L})}, \quad \Delta\beta \neq 0 \qquad (7.1)$$

where

$$H_C = \sum_{i\in C}\left[\frac{p_i^2}{2m_i} + \frac{1}{2}\sum_j \phi_{i,j}\right], \quad C = L, R \qquad (7.2)$$

where L, R denote the left or right bodies in Maxwell's statement of the zeroth "Law", and $\phi_{i,j}$ is the potential energy of particles i, j. Clearly, $H_0(\mathbf{\Gamma}) = H_L(\mathbf{\Gamma}) + H_R(\mathbf{\Gamma})$ and, in contradistinction to the common notation, the interaction energy *between* the two bodies is accounted for within the two sub-Hamiltonians of each body. As usual the energy delta function is taken to be a limitingly thin energy shell, (E, E+dE). We have assumed that there are only pair interactions. We could extend the theory to include many-body interactions, but this would only increase the complexity of the argument without revealing any more physics.

If there were no interactions between the two subsystems L, R, (i.e., $\phi_{i,j} = 0, \ \forall i \in L, j \in R$), these two parts would remain in separate microcanonical equilibriums indefinitely. Such a system would not as a whole be T-mixing. Switching on the interactions between the two subsystems means that the initial system is *not* in thermodynamic equilibrium and, on average, generates future states with positive time-averaged dissipation function.

From Eqs. (7.1) and (7.2), the deviation function is

$$g(\mathbf{\Gamma}) = -\Delta\beta H_L(\mathbf{\Gamma}) \quad \text{or} \quad g(\mathbf{\Gamma}) = \Delta\beta H_R(\mathbf{\Gamma}) \qquad (7.3)$$

If the reciprocal difference is zero, the system is isothermal and is at equilibrium. So, Eqs. (7.1) and (7.2) provide a convenient mathematical model to study thermal relaxation and hence give a proof of Maxwell's zeroth "Law."

The two quantities in Eq. (7.3) differ by a constant $\Delta\beta H_0$, but this constant has no physical relevance and is removed in Eq. (7.1) due to the normalization. The instantaneous dissipation function $\dot{g}(\mathbf{\Gamma})$ is easily seen to be

$$\Omega(\mathbf{\Gamma}) = -\Delta\beta \dot{H}_L(\mathbf{\Gamma}) = \Delta\beta \dot{H}_R(\mathbf{\Gamma}) \qquad (7.4)$$

In deriving this equation, we have used the fact that the energy $H_0 = H_L + H_R$ is a constant of the motion. From the dissipation theorem (Chapter 4), the time evolution of the phase space distribution function is given by

$$f(\mathbf{\Gamma};t) = f(\mathbf{\Gamma};0)\exp\left(-\int_0^{-t} ds\,\Omega(S^s\mathbf{\Gamma})\right) \qquad (7.5)$$

The second law inequality shows that the time integral of the dissipation function satisfies the inequality

$$\left\langle \Omega_t(\mathbf{\Gamma})\right\rangle_0 = -\left\langle \Delta\beta\Delta H_L(S^t\mathbf{\Gamma})\right\rangle_0 = \left\langle \Delta\beta\Delta H_R(S^t\mathbf{\Gamma})\right\rangle_0 > 0, \quad \forall t > 0 \qquad (7.6)$$

where $\Delta H_L(S^t\mathbf{\Gamma}) \equiv H_L(S^t\mathbf{\Gamma}) - H_L(\mathbf{\Gamma})$. If the left side of the system is hotter than the right, that is, $T_L > T_R$, and $\Delta\beta > 0$, in order for Eq. (7.6) to be satisfied, $\langle \Delta H_L(S^t\mathbf{\Gamma})\rangle_0$ must be negative, meaning that the hot left-hand side loses heat energy to the cooler right-hand side, that is, $\langle \Delta H_R(S^t\mathbf{\Gamma})\rangle > 0$. This is in accordance with the second "Law" of thermodynamics.

Substituting into the dissipation theorem for averages gives

$$\langle H_L(S^t\mathbf{\Gamma})\rangle_0 = \langle H_L(\mathbf{\Gamma})\rangle_0 - \Delta\beta \int_0^t ds \langle \dot{H}_L(\mathbf{\Gamma}) H_L(S^s\mathbf{\Gamma})\rangle_0 \tag{7.7}$$

Since the system is T-mixing, in the long-time limit the integral of the correlation function converges to a finite value and the average energy of the left and right sides of the system becomes constant in time. If the average energy of the left hand body is constant, the instantaneous dissipation must have a zero average value:

$$\lim_{t\to\infty} \Delta\beta \langle \dot{H}_L(S^t\mathbf{\Gamma})\rangle_0 = 0 \tag{7.8}$$

In the long-time limit, the average dissipation is zero. This can happen only if the distribution has relaxed to its unique dissipationless equilibrium distribution (Eq. (7.1) with $\Delta\beta = 0$). This is because, as we have already seen, *any* deviation from this distribution must produce a positive value for the time integral of the ensemble-averaged dissipation function (Section 5.3). The temperatures of the left- and right-hand sides of the system must be equal because the unique equilibrium distribution (Eq. (7.1) with $\Delta\beta = 0$) is spatially homogeneous. This completes our proof of the zeroth "Law" of thermodynamics.

Since there is no flux of particles between the two regions or bodies, and if the boundary between the two bodies has a cross-sectional area σ_A, the energy change is simply related to a heat flux $J_{Q,R}$ from the appropriate side of the system

$$\dot{H}_L \equiv J_{Q,R}\sigma_A \tag{7.9}$$

(outward normal convention is used) and we can write the dissipation theorem for the heat flux as

$$\langle J_{Q,R}(S^t\mathbf{\Gamma})\rangle_0 = -\Delta\beta\sigma_A \int_0^t ds \langle J_{Q,R}(\mathbf{\Gamma}) J_{Q,R}(S^t\mathbf{\Gamma})\rangle_0 \tag{7.10}$$

Because of the form of Eq. (7.1), which ensures that all functions that are odd in the momentum (including all fluxes of non-conserved quantities) are zero at time zero, we see that $\langle J_{Q,R}(\mathbf{\Gamma})\rangle_0 = 0$. Equation (7.9) is obviously a form of Fourier's "Law" for heat flow. In fact, it gives an *exact* expression for the nonlinear far-from-equilibrium thermal conductivity (Evans, Williams, and Rondoni, 2012). Indeed, the magnitude of $\Delta\beta$ has not been specified in our derivation and can be arbitrarily large or small. Fourier's "Law" for heat flow only relates to the linear response regime close to equilibrium. It is a linear constitutive relation. Previous time correlation expressions for the thermal conductivity (Evans and Morriss, 1990; Zwanzig, 2001) were also limited to the linear response regime close to equilibrium. In Eq. (7.9), the transient time correlation integral is dependent on the size of the temperature gradient (difference). In the linear response regime, we take the weak gradient limit of the transient time correlation function (TTCF), namely the corresponding *equilibrium* time correlation function.

If the temperature difference is converted into a temperature gradient, and if the heat capacity of the systems is large relative to the heat fluxes, this equation gives an expression for the nonlinear thermal conductivity $J_{Q,L} \equiv \lambda(\Delta\beta)\Delta\beta$ of the pseudo-steady state that may develop initially:

$$\lambda(\Delta\beta) = \sigma_A \int_0^{t_c} ds \langle J_Q(\mathbf{\Gamma}) J_Q(S^t\mathbf{\Gamma}) \rangle_0 \tag{7.11}$$

where t_c is the convergence time for the pseudo-steady state. In the weak gradient limit, Eq. (7.11) is consistent with the Green–Kubo relations for thermal conductivity. (Note that in our system the heat flux appearing in our correlation functions is defined in terms of the energy flux across a plane whereas the usual heat flux appearing in Green–Kubo expressions is defined over a (homogeneous) volume (Evans and Morriss, 1990). Also, J_Q appearing in Eq. (7.11) can be either the left or right fluxes as the formula is symmetric.) However, for our system, because the total energy is conserved but the energies of each of the two regions are not separately conserved, and our T-mixing system eventually relaxes to equilibrium, the heat flux eventually goes to zero:

$$\lim_{t\to\infty} \langle J_{Q,R}(S^t\mathbf{\Gamma}) \rangle_0 = -\Delta\beta\sigma_A \int_0^{\infty} ds \langle J_{Q,R}(\mathbf{\Gamma}) J_{Q,R}(S^t\mathbf{\Gamma}) \rangle_0 = 0 \tag{7.12}$$

In this equation, all dynamics is Newtonian and the initial distribution is the initial nonequilibrium distribution Eq. (7.1). It is valid when the initial state is arbitrarily far from equilibrium. In the far-from-equilibrium regime, the time correlation function is not an equilibrium time correlation function but rather is a TTCF (nonequilibrium) that is dependent on the magnitude of the initial temperature difference.

Equation (7.11) is a special case of the heat death equation (Eq. (5.36)). Infinite-time correlation integrals of fluxes of nonconserved quantities vanish.

7.3 Steady-State Heat Flow (Searles and Evans, 2001; Evans, Searles, and Williams, 2010)

In previous chapters we have mostly considered mechanically driven systems where the dissipative field appears explicitly in the equations of motion as per Eq. (6.4). In the previous section we studied transient relaxation due to heat flow in a field-free Hamiltonian system. Here, because of the importance of understanding thermal conductivity, we will give a method for studying steady-state boundary-driven heat flow (Searles and Evans, 2001; Evans, Searles, and Williams, 2010).

Experimentally, there are a number of ways in which walls can be thermostatted. If the walls are made of highly thermally conductive material, chilled and hot fluids may be circulated through pipes in the walls while having their relative proportions adjusted according to some temperature-sensing feedback mechanism. This is essentially the same as what we accomplish mathematically in the differential feedback isokinetic thermostat or in the integral feedback mechanism of the Nosé–Hoover thermostat. Alternatively, if the heat capacity of the reservoirs is huge compared to that of the thermal conduction cell, then the temperature variation in the reservoirs over microscopic relaxation times may be regarded as insignificant.

Here we employ the so-called Nosé–Hoover thermostat in the reservoir regions in order to maintain these regions at a fixed temperature. Its impact on the system of interest, namely the thermal conduction cell, is only indirect. In an experiment, the material properties of the thermal conduction cell are independent of whether the reservoirs are maintained at a fixed temperature by virtue of the circulation of a coolant or the use of reservoirs of large heat capacity. The thermal conductivity is a material property. It is independent of the precise chemical composition of the walls of a conduction cell.

The theory that follows is also independent of the thermostatting mechanism. The reason for this independence is that the formal fluctuation formulae are independent of precisely how the thermostatting is accomplished in the far-removed thermostatting region. We can move the thermostatting region arbitrarily far from the system of interest and still generate the same fluctuation relation. There is no way that the system of interest can "know" precisely how the heat is ultimately removed by the remote thermostat. We note that in low-dimensional anharmonic chains it is well known that there can be long-range spatial correlations for heat flow – see Gallavotti (2008). In typical physical systems, such correlations are much shorter ranged.

It turns out that the only significant thing that the system of interest takes from the thermostat is the equilibrium thermodynamic temperature the system will relax to, if it is so allowed.

The aim is to derive formulae for the dissipation function of a thermal conduction cell (Searles and Evans, 2001). We consider the system initially at equilibrium (because then the phase space distribution function is known – see Chapter 5). Initially the whole system is isothermal. The temperature gradient is then applied and a heat flux develops in time.

The equations of motion for all the particles in the combined systems, $H, 0, C$ are

$$\begin{aligned} \dot{\mathbf{q}}_i &= \mathbf{p}_i \\ \dot{\mathbf{p}}_i &= \mathbf{F}_i - \alpha_H \mathbf{p}_i A_i - \alpha_C \mathbf{p}_i B_i \end{aligned} \tag{7.13}$$

where $\alpha_{H/C}$ are the thermostat multipliers, $T_{H/C}$ are the target temperatures of the hot and cold regions, and A_i and B_i are switches equal to 1 or 0. A_i is only one for particles in region H, and B_i is only one for particles in region C. The multipliers themselves satisfy the following equations of motion:

$$\frac{d\alpha_{H/C}}{dt} = \frac{1}{\xi}\left(\sum_{i\in H/C} \frac{\mathbf{p}_i^2}{m} - \left(3N_{\text{th}} + 1\right) k_B T_{H/C}\right) \tag{7.14}$$

where N_{th} is the number of particles in each reservoir. The constant ξ controls the timescale for fluctuations in the kinetic temperatures of regions H,C. The time constant is given by (Evans and Morriss, 1990) $\tau_{H/C} = O(\sqrt{\xi/3N_{\text{th}}k_{\text{B}}T_{H/C}})$. We always choose $\xi = O(N_{\text{th}})$ so that $\tau_{H/C}$ is intensive.

For simplicity, we assume that the walls are sufficiently dense so that the particles from region 0 do not penetrate either of the reservoir regions – the walls are impervious solids. The details of the interatomic forces implicit in $\{\mathbf{F}_i\}$ will be described later. It is important to note that in the zero region and the $H, 0$ and $C, 0$ interfaces, the equations of motion can be made arbitrarily realistic by improved modeling of the interatomic forces. In the zero region, there are no unnatural forces. Our system is very similar to that studied by Searles and Evans (2001) and Petravic and Harrowell (2006, 2005), and although the dimensionality and the particle dynamics is very different, it has the same form for the dissipation function as in the system studied by Mejia-Monasterio and Rondoni (2008).

The additional Nosé–Hoover thermostat ensures that in a steady state the reservoir regions are maintained at constant time-averaged kinetic temperatures T_H, T_C. In a nonequilibrium T-mixing, steady state

$$\lim_{t\to\infty} d\overline{\alpha}_{H/C,t}/dt = 0 \Rightarrow 1/(3N_{\text{th}} + 1)k_{\text{B}} \sum_{i\in H/C} \frac{\overline{\mathbf{p}_{i,t}^2}}{m} = T_{H/C} \tag{7.15}$$

Since the system is assumed to be T-mixing, it must be physically ergodic. We assume that at $t = 0$ the initial phase space distribution $f(\boldsymbol{\Gamma}; 0)$ is the equilibrium canonical distribution (Section 5.4):

$$f(\boldsymbol{\Gamma}^\dagger; 0) = \frac{\exp\left[-\beta_0\left(H_0(\boldsymbol{\Gamma}) + \xi(\alpha_H^2 + \alpha_C^2)/2\right)\right]}{\int d\boldsymbol{\Gamma}' \exp\left[-\beta_0\left(H_0(\boldsymbol{\Gamma}) + \xi(\alpha_H^2 + \alpha_C^2)/2\right)\right]} \tag{7.16}$$

where $\beta_0 = 1/(k_{\text{B}}T_0)$, $H_0 = \sum p_i^2/2m + \Phi(q)$ is the internal energy, $H_0^\dagger = H_0 + \xi(\alpha_H^2 + \alpha_C^2)/2$ is called the *extended energy*, and $\boldsymbol{\Gamma}^\dagger \equiv (\boldsymbol{\Gamma}, \alpha_H, \alpha_C)$ is the extended phase space vector.

The phase space expansion factor $\Lambda(\boldsymbol{\Gamma}^\dagger)$ appearing in the phase continuity equation is

$$\Lambda = -dN_{\text{th}}\alpha_H - dN_{\text{th}}\alpha_C \tag{7.17}$$

Thus the formal Lagrangian solution of the phase continuity equation is – see Eq. (2.19)

$$f(S^t\boldsymbol{\Gamma}^\dagger; t) = f(\boldsymbol{\Gamma}^\dagger; 0)\exp\left[\int_0^t ds\, dN_{\text{th}}\alpha_H(s) + dN_{\text{th}}\alpha_C(s)\right] \tag{7.18}$$

From the equations of motion, we see that the rate of change of the extended internal energy is

$$\dot{H}_0^\dagger = -3N_{\text{th}}k_{\text{B}}(T_H\alpha_H + T_C\alpha_C) \tag{7.19}$$

Substituting Eqs. (7.17) and (7.19) into Eq. (3.2) gives the time-averaged dissipation function as

$$\begin{aligned}\overline{\Omega}_t(\Gamma^\dagger)t &= -\int_0^t ds\left[3N_{\text{th}}\beta_0 k_B\left(T_H\alpha_H(s) + T_C\alpha_C(s)\right) - 3N_{\text{th}}(\alpha_H(s) + \alpha_C(s))\right]\\ &= \int_0^t ds 3N_{\text{th}}\{\alpha_H(s)(T_C - T_H) + \alpha_C(s)(T_H - T_C)\}/(T_H + T_C)\\ &= 3N_{\text{th}}\frac{(T_H - T_C)}{(T_H + T_C)}\int_0^t ds\,[\alpha_C(s) - \alpha_H(s)] \end{aligned} \tag{7.20}$$

where the second line follows from the imposed relationship between the initial temperature and target temperatures of the thermostats, $2T_0 = T_C + T_H$. From the Evans–Searles fluctuation theorem (ESFT) (Chapter 3), we see that the probability ratio of observing conjugate values for the time-averaged difference in the thermostat multipliers is

$$\frac{p(\overline{\alpha}_{C,t} - \overline{\alpha}_{H,t} = A)}{p(\overline{\alpha}_{C,t} - \overline{\alpha}_{H,t} = -A)} = \exp\left[3N_{\text{th}}\frac{T_H - T_C}{T_C + T_H}At\right] \tag{7.21}$$

The ESFT for heat flow given by Eq. (7.21) is exact for any arbitrary system size, observation time t, and also arbitrarily far from equilibrium.

Equation (7.21) is a statement of the transient fluctuation theorem for heat flow between Nosé–Hoover thermostatted walls. Since the system is T-mixing, the system will relax to a unique steady state. Therefore we can consider the steady-state fluctuation theorem

$$\lim_{t\to\infty} \ln\left[\frac{p\left(\overline{\alpha}_{C,t} - \overline{\alpha}_{H,t} = A\right)}{p(\overline{\alpha}_{C,t} - \overline{\alpha}_{H,t} = -A)}\right]\Bigg/\left[3N_{\text{th}}\frac{T_H - T_C}{T_C + T_H}t\right] = A \tag{7.22}$$

These two equations, that is, Eqs. (7.21) and (7.22), are valid outside the linear regime. For our thermal conduction setup and with our initial conditions and thermostats, the only caveat is that the steady-state formula requires the system to be T-mixing. Equations (7.21) and (7.22) are clearly consistent with the second law of thermodynamics in that it is exponentially more probable for heat to flow from hot to cold, in which case $\overline{\alpha}_{C,t} > 0$, $\overline{\alpha}_{H,t} < 0$ and from (7.22) we see that in the steady state $\lim_{t\to\infty}|\overline{\alpha}_{C,t}| > \lim_{t\to\infty}|\overline{\alpha}_{H,t}|$. In either the large system and/or the long-time limit, the time-averaged heat will flow only from hot to cold.

The dissipation theorem (Chapter 4) gives an exact TTCF expression for the ensemble average of the nonlinear response of an arbitrary phase variable $B(\Gamma)$ as

$$\langle B(t)\rangle = \langle B(0)\rangle - \frac{3N_{\text{th}}(T_H - T_C)}{T_H + T_C}\int_0^t ds\langle B(s)[\alpha_H(0) - \alpha_C(0)]\rangle \tag{7.23}$$

In this equation, the angle brackets denote an average over the initial (i.e., $t = 0$) ensemble and $\langle B(t)\rangle \equiv \langle B(S^t\Gamma)\rangle$. Unlike the fluctuation theorems, the dissipation theorem does *not* require ergodicity. The linearized weak-field version of this equation is essentially identical to that in the paper by Petravic and Harrowell (2006, 2005, Eq. (7.23)). By comparing with the usual Kawasaki distribution function for a system driven by an external mechanical force, we see that, although the system is a *thermal* nonequilibrium system where boundary conditions rather than external mechanical forces drive the system away from equilibrium, there is a *formal* resemblance of the nonlinear response to that obtained if we applied a fictitious mechanical field

$$F_e = \frac{k_B(T_H - T_C)}{2} \tag{7.24}$$

to the system. In this case, the intensive dissipative flux J can be identified as the fictitious function

$$J(\mathbf{\Gamma}^\dagger) = 3n_{\text{th}}k_B[\alpha_H(\mathbf{\Gamma}^\dagger) - \alpha_C(\mathbf{\Gamma}^\dagger)] \tag{7.25}$$

where $n_{\text{th}} \equiv {}^{N_{\text{th}}}/_{V_{\text{th}}}$ is the number density of the thermostat volumes.

Equation (7.23) contains a great deal of information. Since the system is T-mixing, the TTCF appearing on the right-hand side decays to zero at long times. More precisely, we require that time integrals of TTCFs of phase functions and the time zero dissipation function (7.23) should converge in the infinite-time limit.

We note that in the weak field (see Eq. (7.23)) limit the linear response of the system to thermal conduction can be computed *exactly* from the time integral of an equilibrium time correlation function. In this limit, T-mixing implies that the equilibrium dynamics is mixing and that the correlations decay faster than $1/t$.

Choosing the phase function $B(\mathbf{\Gamma})$ to be the dissipation function itself and using the second law inequality shows that at late times the *ensemble-averaged* dissipation function equals the *time-averaged* dissipation function and that the average value is nonnegative. In fact, it must be strictly positive because the equilibrium state is the unique dissipationless state for T-mixing systems.

If the transient autocorrelation function for the dissipation function is positive for all times, then the relaxation to the steady state is monotonic and the steady state corresponds to the state of *maximum* dissipation compared to all the transient states, the equilibrium state, and, of course, all the conjugate time-reversed antistates. If the autocorrelation function is not positive for all time, then the steady state has no such extremal properties (i.e., there are transient states with greater ensemble-averaged dissipation than the steady state).

Our system considers the transient response of the three regions H, 0, C that are initially at the same temperature T_0. At $t = 0$, systems H and C are instantly brought into contact with Nosé–Hoover thermostats, which rapidly $\left[\tau_{H/C} = O\left(\sqrt{\xi/3N_{\text{th}}k_B T_{H/C}}\right)\right]$ bring systems H and C to temperatures T_H, T_C, respectively.

Without loss of generality, we assume that the three regions H, 0, and C have a rectangular cross section of area σ_A and wall normals parallel to the x-axis, and the distance separating the thermostatted reservoirs is L.

At the moment, our expression for the dissipation function involves nonphysical variables. We now transform our expression for the dissipation function into an expression involving physically measureable variables: heat fluxes and temperature difference.

The ultimate fluxes into and out of our system are given by the energy gain or loss by the thermostats themselves. These are the only nonconservative elements of our system. The thermostatting terms are analogous to the coolant in a physical thermostat. Once the energy is taken up by the circulating coolant, the physical circulation of that coolant removes that energy from the system of interest. With this in mind, it is natural to evaluate the H/C "heat fluxes" as

$$J_{QH/C}(t)\sigma = \mp 3N_{\mathrm{th}}k_{\mathrm{B}}T_{H/C}\alpha_{H/C}(t) \tag{7.26}$$

These are the heat fluxes across the area σ_A *immediately* before, or after, the heat is removed, or injected, by the thermostats themselves. (Note, the difference in signs! $\lim_{t\to\infty}\overline{\alpha}_{H,t} = -\lim_{t\to\infty}\overline{\alpha}_{C,t} < 0$. This further implies: $\lim_{t\to\infty}\overline{J}_{QH,t} = \lim_{t\to\infty}\overline{J}_{QC,t} > 0$.)

We begin by manipulating our expression for the instantaneous dissipation function. From Eqs. (7.20) and (7.26), we see that

$$\begin{aligned}\Omega(t)k_{\mathrm{B}} &= -\frac{1}{T_0}\left[3N_{\mathrm{th}}k_{\mathrm{B}}T_H\alpha_H(t) - 3N_{\mathrm{th}}k_{\mathrm{B}}T_0\alpha_H(t)\right.\\ &\quad\left.+3N_{\mathrm{th}}k_{\mathrm{B}}T_C\alpha_C(t) - 3N_{\mathrm{th}}k_{\mathrm{B}}T_0\alpha_C(t)\right]\\ &= \left[-\frac{J_{QH}(t)\sigma_A}{T_H} + \frac{J_{QC}(t)\sigma_A}{T_C}\right] + \left[\frac{J_{QH}(t)\sigma_A}{T_0} - \frac{J_{QC}(t)\sigma_A}{T_0}\right]\end{aligned} \tag{7.27}$$

Equation (7.27) consists of a sum of two terms.

Definition

The second term is not explicitly proportional to a function of the temperature difference – it is not a *driven* term but rather is a *boundary term*. This second term cancels another boundary term inherent in the first term and ensures that if the system is in true thermodynamic equilibrium where $T_H = T_C = T_0$, the dissipation function is identically zero in spite of the fact that due to boundary fluctuations $J_H(t) \neq J_C(t)$, instantaneously.

At equilibrium, it is only on average that these two quantities are equal. If the second term is missing from Eq. (7.28), the relative importance of these boundary fluctuations increases without bound as the system becomes closer to equilibrium.

This expression for the dissipation function involves only physically measurable variables that retain their physical meaning even when the system of interest is arbitrarily far from equilibrium. Because there is no convection in our system, the heat fluxes are simply energy fluxes. The target temperatures of the artificial Nosé–Hoover thermostat, $T_{H/C}$, are the equilibrium thermodynamic temperatures that the two thermostats would relax to if the two thermostatting regions were decoupled from each other and the system of interest, and each thermostat was allowed to relax to equilibrium – see Chapter 5.

Substituting Eq. (7.27) into the second law inequality gives

$$\begin{aligned}\lim_{t\to\infty} k_{\mathrm{B}}\overline{\Omega}_t &= \lim_{t\to\infty}\left[\frac{\overline{J}_{Qt}\sigma_A}{T_C} - \frac{\overline{J}_{Qt}\sigma_A}{T_H} + \frac{\overline{J}_{Qt}\sigma_A}{T_0} - \frac{\overline{J}_{Qt}\sigma_A}{T_0}\right]\\ &= \lim_{t\to\infty}\overline{J}_{Qt}\sigma_A\left[\frac{1}{T_C} - \frac{1}{T_H}\right] > 0\end{aligned} \tag{7.28}$$

For our system this inequality is clearly satisfied. It is a product of two positive terms.

Aside

For those who are familiar with linear irreversible thermodynamics, we can see that *in the local equilibrium regime close to and only close to equilibrium,* the time-averaged dissipation function in Eq. (7.27) is recognizable as the extensive spontaneous entropy production Σ_{therm} (De Groot and Mazur, 1984):

$$\begin{aligned}
\Sigma_{\text{therm}}(t) &= \lim_{\nabla T\to 0}\int_V d\mathbf{r}\ \sigma(\mathbf{r},t) = \lim_{\nabla T\to 0}\int_V d\mathbf{r}\ \mathbf{J}_Q(t)\cdot\nabla T^{-1} \\
&= \sigma_A \lim_{\nabla T\to 0}\int_{-L/2}^{+L/2} dx\ J_{Qx}(t)\frac{d(1/T(x))}{dx} \\
&= \sigma_A \left.\frac{J_{Qx}(x,t)}{T(x)}\right|_{-L/2}^{+L/2} - \sigma_A\int_{-L/2}^{+L/2} dx\ \frac{1}{T(x)}\frac{dJ_{Qx}(x,t)}{dx} \\
&= \lim_{T_H-T_C\to 0}\left(-\sigma_A\left(\frac{J_{QH}(t)}{T_H}-\frac{J_{QC}(t)}{T_C}\right)\right. \\
&\quad \left. +\sigma_A\frac{J_{QH}(t)-J_{QC}(t)}{T_0}+O\left(\frac{d^3}{dx^3}\right)\right)
\end{aligned} \tag{7.29}$$

This expression for the thermodynamic entropy production, $\Sigma_{\text{therm}}(t)$, equals the average dissipation function multiplied by Boltzmann's constant: $k_B\Omega(t)$ in Eq. (7.27). However, unlike the entropy production, the definition of the dissipation function retains precise mathematical meaning far from equilibrium. In a sense, therefore, the dissipation function serves as a mathematical replacement for the entropy production. When the entropy production can be defined, it is equal, on average, to the dissipation function. However, unlike entropy production, the dissipation function can, for ergodically consistent systems, always be defined.

Aside

It is a trivial matter to compute the time derivative of the fine-grained Gibbs entropy Eq. (5.53), $S_G = -k_B\int d\mathbf{\Gamma}^\dagger\ f(\mathbf{\Gamma}^\dagger,t)\ln(f(\mathbf{\Gamma}^\dagger,t))$. From Eq. (6.9) we see that

$$\begin{aligned}
\dot{S}_G(t) &= k_B\int d\mathbf{\Gamma}^\dagger\ f(\mathbf{\Gamma}^\dagger;t)\,\partial\dot{\mathbf{\Gamma}}^\dagger(t)/\partial\mathbf{\Gamma}^\dagger \\
&= -3N_T k_B\langle\alpha_H(t)+\alpha_C(t)\rangle = k_B\langle\Lambda(t)\rangle
\end{aligned} \tag{7.30}$$

It is clear that the dissipation function is *not* instantaneously related to the time derivative of the fine-grained Gibbs entropy.

Comparing Eq. (7.27) with Eq. (7.30), we see that $\dot{S}_G(t) = \langle J_H(t)\rangle\sigma_A/T_H - \langle J_C(t)\rangle\sigma_A/T_C$ does not contain the boundary terms and so (in contrast to the instantaneous dissipation) $\dot{S}_G(t) \neq 0$ for equilibrium systems.

Only the time-averaged rate of change of the fine-grained Gibbs entropy for the steady state is equal, for sufficiently long averaging times, to -1 times the steady-state average of the thermodynamic entropy production:

$$\begin{aligned} -\bar{\dot{S}}_{G,t} &= -k_B\langle\overline{\Lambda}_t\rangle \\ &\underset{t\to\infty}{=} -\sigma_A\langle\bar{J}_{Q,t}\rangle\left(\frac{1}{T_H}-\frac{1}{T_C}\right) \\ &\underset{t\to\infty}{=} k_B\langle\overline{\Omega}_t\rangle \end{aligned} \tag{7.31}$$

The second equality is simply a restatement of Eq. (7.29); the third equality is true for sufficiently long averaging times. We note that $\Lambda(t)$ is the argument of the steady-state Gallavotti–Cohen fluctuation relation – see Section 6.10. The discussion above about the relative importance of boundary terms shows that, as we said in Section 6.10, ever longer averaging times are required for the asymptotic Gallavotti–Cohen fluctuation theorem to converge as we approach equilibrium. As equilibrium is approached, this time grows like $(T_H - T_C)^{-2}$!

Applying the second law inequality, Eq. (3.11), to our system shows that in the long time limit we obtain

$$\begin{aligned} \lim_{t\to\infty}\langle\bar{J}_{Q,t}\rangle &\equiv -\lambda\left({}^{\partial T}/_{\partial x}\right)\frac{\partial T}{\partial x} \\ &= -\lambda\left({}^{\partial T}/_{\partial x}\right)\frac{\partial T}{\partial x} > 0 \Rightarrow \lambda\left({}^{\partial T}/_{\partial x}\right) > 0, \quad \forall\, {}^{\partial T}/_{\partial x} \end{aligned} \tag{7.32}$$

where $\lambda\left({}^{\partial T}/_{\partial x}\right)$ is the nonlinear thermal conductivity, and the temperature gradient is negative. Since the system is T-mixing, a unique steady state is generated for any given temperature gradient and the nonlinear thermal conductivity is finite and positive for any finite temperature gradient. Therefore, Fourier's "Law," which applies to the weak-field limit, is valid for our system. (The thermal conductivity cannot be zero since in T-mixing systems the only state that has zero dissipation is the equilibrium state – see Chapter 6.)

7.4
Dissipation Theorem for a Temperature Quench

In this section we will discuss response theory for a temperature quench (Schmidt and Evans, 1994) and its relationship to the relaxation theorem. The response of a system to a rapid change in temperature is of interest in the study of phase transitions. Such quenches may be used to induce glass formation or to initiate spinodal decomposition. Typically, a quench takes the system from an initial equilibrium state to a final equilibrium state that is characterized by a different value (usually lower) of the temperature. In the case of a quench to a glass, the final state is not in true thermodynamic equilibrium. The transition states between the initial and the final states are also obviously nonequilibrium states. These observations suggest that one should be able to analyze such transitions using a standard tool of nonequilibrium statistical mechanics, namely nonlinear response theory (Schmidt and Evans, 1994). In this section we will show one way in which this can be accomplished. In the next chapter (Chapter 8) we will discuss another set of ways of studying the transitions between equilibrium states via nonequilibrium pathways.

Here we describe equations of motion that give an exact description of an impulsive temperature quench of arbitrary magnitude. We describe the applied field and its conjugate dissipative flux. We apply the dissipation theorem (Chapter 4) to develop an exact TTCF expression, which may be used to analyze the transient response of any phase function to a temperature quench. These expressions have been tested in computer simulations and found to be consistent with the observed response.

In this section we consider an N-particle classical system evolving under Gaussian isokinetic equations of motion – see Eq. (6.4):

$$\dot{\mathbf{q}}_i = \frac{\mathbf{p}_i}{m}$$
$$\dot{\mathbf{p}}_i = \mathbf{F}_i - \alpha\mathbf{p}_i \tag{7.33}$$

where $\mathbf{p}_i$, $\mathbf{q}_i$, and $\mathbf{F}_i$ are the momenta, positions, and interatomic force vectors for the particle i. We assume that the initial total momentum is zero. Note that we could employ our usual switch as in Eq. (6.4) but the mathematics hardly changes.

It has been proved (Evans and Sarman, 1993) that in an equilibrium system that is *mixing*, time averages under Gaussian thermostatted dynamics are identical to the corresponding time averages under Newton's equations, in the thermodynamic limit. In this same limit, equilibrium time correlation functions evaluated under thermostatted or Newtonian dynamics are also equivalent. In Eq. (7.33), the internal energy is given by $H_0 = \sum_i(K_i) + \Phi$, where $K_i = \frac{\mathbf{p}_i^2}{2m}$ is the (peculiar) kinetic energy of particle i, and Φ is the total potential energy. The internal energy and total momentum are constants of the motion.

In order to maintain a constant temperature, we set $\dot{K} \equiv \sum_i \dot{K}_i = 0$ and thus

$$\dot{K} = \sum_i \frac{\mathbf{p}_i}{m}\cdot(\mathbf{F}_i - \alpha\mathbf{p}_i) = 0 \Rightarrow \alpha = \frac{\sum_i \frac{\mathbf{p}_i}{m}\cdot\mathbf{F}_i}{2K} \tag{7.34}$$

We now consider a system that obeys the above equations of motion, Eqs. (7.33) and (7.34), and experiences an instantaneous change in its kinetic temperature at time $t = 0$. The equilibrium distribution function immediately before the quench is isokinetic – see Section 6.2:

$$f(\boldsymbol{\Gamma}; 0^-) = \frac{\exp[-\beta_\_\Phi(\boldsymbol{\Gamma})]\delta(\mathbf{p})\delta(2K - (3N-4)k_B T(0^-))}{\int_D d\boldsymbol{\Gamma}\exp[-\beta_\_\Phi(\boldsymbol{\Gamma})]\delta(\mathbf{p})\delta(2K - (3N-4)k_B T(0^-))} = f_K(\boldsymbol{\Gamma}; \beta_\_) \tag{7.35}$$

with $\beta_\_ = 1/k_B T(0^-)$, which is a function of the equilibrium temperature $T(0^-)$ for times $t \leq 0^-$. $\boldsymbol{\Gamma} = (x_1, x_2, \ldots, z_N, p_{x1}, \ldots, p_{zN})$ is the $6N$-dimensional phase space vector.

At time $t = 0$, a temperature quench is accomplished by scaling all momenta by a factor λ while the coordinates remain unchanged; that is, $\mathbf{p}_i(0^+) = \lambda\mathbf{p}_i(0^-)$ and $\mathbf{q}_i(0^+) = \mathbf{q}_i(0^-)$. With this scaling, the kinetic energy and kinetic temperature $T \equiv \sum \mathbf{p}_i^2/3Nmk_B$ change, but the potential energy, which is a function of the coordinates alone, does not. The total momentum remains at zero after the rescaling. The distribution function immediately after the quench $t = 0^+$ is therefore given by

$$f(\mathbf{\Gamma}, 0^+) = \frac{\exp[-\beta_-\Phi(\mathbf{\Gamma})]\delta(\mathbf{p})\delta(2K - (3N-4)k_B T_+)}{\int_D d\mathbf{\Gamma} \exp[-\beta_-\Phi(\mathbf{\Gamma})]\delta(\mathbf{p})\delta(2K - (3N-4)k_B T_+)} \tag{7.36}$$

where $\beta_+ = \beta_-/\lambda^2 = 1/k_B T_+ = 1/k_B T_\infty$.

Clearly, $f(\mathbf{\Gamma}; 0^+)$ is not an equilibrium distribution function. The unique isokinetic equilibrium distribution function is given by Eq. (7.35) with $\beta_- \to \beta_+$. However, it is a straightforward matter to compute the dissipation function so that we can study the time evolution and relaxation of this system.

From Eq. (7.36) and the definition of the instantaneous dissipation function, Eq. (3.7), we get

$$\Omega = \beta_-\dot{\Phi} + (3N-4)\alpha \tag{7.37}$$

Using the fact that

$$\dot{H}_0(t) = -2K_+\alpha(t) = -(3N-4)k_B T_+\alpha(t) \equiv -\dot{Q}(t) \tag{7.38}$$

where $\dot{Q}$ is the heat absorbed from the system by the thermostat

$$\Omega(\mathbf{\Gamma}) = \left(1 - {}^{\beta_-}/_{\beta_+}\right)(3N-4)\alpha(\mathbf{\Gamma}) = (\beta_+ - \beta_-)\dot{Q} \tag{7.39}$$

If the quench involves a sudden cooling, we would expect that $\beta_+ > \beta_-$ and this equation is in accordance with what we expect from the second law inequality (i.e., $\langle\dot{Q}\rangle > 0$ for a temperature *quench*). The configurational degrees of freedom are still "hot" (see Eq. (5.55)) and, in order for the thermostat to maintain the kinetic degrees of freedom at the suddenly lower temperature, the thermostat must remove energy from the system (i.e., $\langle\alpha\rangle > 0$) until the configurational degrees of freedom have also "cooled down."

If we substitute Eq. (7.39) into the dissipation theorem Eq. (4.7), we see that the time dependence of the ensemble average of a phase function $A(\mathbf{\Gamma})$ can be computed as

$$\langle A(t)\rangle = \langle A(0^+)\rangle + (\beta_+ - \beta_-)\int_{0^+}^{t} ds \langle\dot{Q}(0^+)A(s)\rangle. \tag{7.40}$$

If we set $A(\mathbf{\Gamma})$ to be $\dot{Q}(t)$, we find that

$$\begin{aligned}\langle\dot{Q}(t)\rangle &= \langle\dot{Q}(0^+)\rangle + (\beta_+ - \beta_-)\int_{0^+}^{t} ds\langle\dot{Q}(0^+)\dot{Q}(s)\rangle \\ &= (\beta_+ - \beta_-)\int_{0^+}^{t} ds\langle\dot{Q}(0^+)\dot{Q}(s)\rangle\end{aligned} \tag{7.41}$$

We know from the relaxation theorem that, if the system is T-mixing, at long times the system will relax to an isokinetic distribution at a temperature $T_\infty = T_+$. This equation predicts the direct response of an arbitrary phase function $A(\mathbf{\Gamma})$ to an instantaneous quench in temperature. It also shows that for temperature quenches the relative difference in the initial and final temperatures, $(\beta_- - \beta_\infty)/\beta_\infty = T_\infty/T_- - 1$, plays the role of the "applied external field" in response theory. The dissipative flux that is conjugate to this "field" is simply $\beta_\infty \dot{Q}(t)$. Although the quench is initiated by an impulsive action (i.e., the momentum scaling at $t = 0$), the "external field" in these equations takes the form of a Heaviside step rather than a delta function in time.

Aside

This problem is impossible to tackle using conventional linear, irreversible thermodynamics. The initial quenched state is just too far from equilibrium for conventional thermodynamics to say anything about the problem.

7.5 Color Relaxation in Color Blind Hamiltonian Systems

It is frequently useful in nonequilibrium statistical mechanics to endow otherwise identical particles with a color label. One can invent a fictitious color field that interacts with the color to induce color currents or to set up stationary gradients in color density in a fluid. This can provide useful information regarding self-diffusion in fluids – see Evans *et al.* (1983) and Evans and Morriss (1990, Section 6.2).

We now consider the free relaxation of a color density modulation (Evans, Searles, and Mittag, 2001; Evans and Searles, 2002). First, we need to construct an ensemble of systems with a color density modulation. Without loss of generality, consider a system of N identical particles, which for $t < 0$ is subject to a color Hamiltonian

$$H_c = H_0 + F_c \sum_{i=1}^{N} c_i \sin(kx_i) \tag{7.42}$$

where $c_i = (-1)^i$ is the color charge of particle i, $k = 2\pi/L$, where L is the boxlength, and $H_0 \equiv \sum_i p_i^2/2m + \in \sum_{i<j} \Phi(\mathbf{q})$ is a "color blind" interaction Hamiltonian (the potential energy $\Phi(\mathbf{q})$ and the internal energy $H_0(\mathbf{\Gamma})$ do not refer to the color charges). The color density modulation can be measured by averaging the appropriate Fourier component:

$$\rho_c(k) \equiv \sum_{i=1}^{N} c_i \sin(kx_i) \tag{7.43}$$

where x_i is the x-coordinate of particle i. We assume that for $t < 0$, the system is in contact with a heat bath. Since the system is at thermal equilibrium for $t < 0$, the color field induces a color density wave

$$\begin{aligned} \langle \rho_c(k,0) \rangle_{F_c} &= \frac{\int d\mathbf{\Gamma}\ \rho_c(k) \exp[-\beta(H_0 + F_c \rho_c(k))]}{\int d\mathbf{\Gamma} \exp[-\beta(H_0 + F_c \rho_c(k))]} \\ &\overset{F_c \to 0}{=} -\beta F_c \langle \rho_c(k)^2 \rangle_{F_c = 0} \end{aligned} \tag{7.44}$$

From the last line of Eq. (7.44), it is clear that in the weak-field limit $\lim_{F_c \to 0} \langle \rho_c(k,0) \rangle_{F_c = 0} < 0$. So at $t = 0$ the system is initially modulated with a color density wave. We wish to consider the behavior of the system for $t \geq 0$, when the external color field is "turned off" and the contact between the system and the heat bath is broken. The system then relaxes freely under the "color blind" Hamiltonian H_0.

For $t > 0$, no work is done on the system and no heat is transferred to or from the system's surroundings. The system evolves with constant energy ($E = H_0(\mathbf{\Gamma})$). The fine-grained Gibbs entropy is constant, Eq. (2.56). No calorimetric process can reveal a change in the entropy. Indeed, in the linear response regime in the weak-field limit (and only in this limit!), if we disregard the color labels, the system would be thought of as being in equilibrium. Nevertheless, according to Le Chatelier's principle, the color density modulation should decay rather than grow as the system becomes homogeneous with respect to color.

The dissipation function $\Omega(\Gamma)$ can be determined using Eq. (3.2). For $t > 0$, there is no phase space compression since the dynamics is Newtonian and there is no applied field. Furthermore, energy is conserved. Therefore, the dissipation function becomes

$$\begin{aligned} t\overline{\Omega}_t &= \beta[H_c(t) - H_c(0)] \\ &= \beta F_c \int_0^t ds\, \dot{\rho}_c(k,s) = \beta F_c[\rho_c(k,t) - \rho_c(k,0)], \quad \forall t > 0 \end{aligned} \tag{7.45}$$

The time-integrated dissipation function thus gives a direct measurement of the change in the color density modulation order parameter. This is an interesting system to study, and it emphasizes that the dissipation function is a functional of *both* the initial distribution of states and the dynamics. In this example, the dynamics is purely Newtonian with no external driving fields or thermostats. It is the form of the initial distribution that generates dissipation. This is in spite of the fact that for $t > 0$ the dynamics is "color blind" and for weak fields, in the linear response regime, if you ignore the color label, the initial distribution would be an equilibrium distribution and the number density of particles would, on average, be uniform.

Applying the ESFT equation, Eq. (3.6), to this system gives

$$\frac{p[\rho_c(k,t) - \rho_c(k,0) = A]}{p[\rho_c(k,t) - \rho_c(k,0) = -A]} = \exp[\beta F_c A] \tag{7.46}$$

where β is the reciprocal temperature of the initial ensemble.

In order to test this equation, we considered a system of 32 particles in 2 Cartesian dimensions. The particles interact via a WCA potential, and the equations of motion at $t < 0$ are

$$\begin{aligned} \dot{\mathbf{q}}_i &= \frac{\mathbf{p}_i}{m}, \quad \forall t \\ \dot{\mathbf{p}}_i &= \mathbf{F}_i - \mathbf{i}c_i kF_c \cos kx_i - \alpha\mathbf{p}_i, \quad t < 0 \\ &= \mathbf{F}_i, \quad t > 0 \\ \dot{\alpha} &= \frac{1}{Q}\left(\sum_{i=1}^{N} \frac{\mathbf{p}_i^2}{m} - 2Nk_B T\right), \quad t < 0 \\ &= 0, \quad t > 0 \end{aligned} \tag{7.47}$$

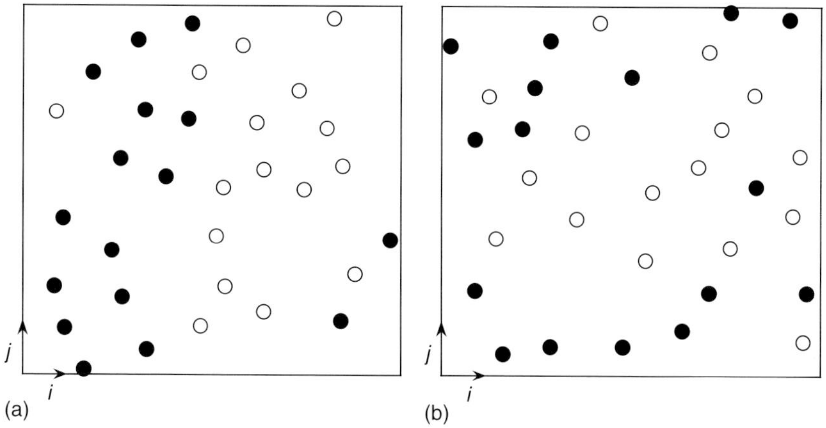

Figure 7.1 Snapshots from a molecular dynamics simulation showing the phase separation of black and white particles at $t<0$ (with field on) in (a) and their relaxation to equilibrium (at $t=32$) in (b). Here, $T=1.0$, $n=0.4$, and $F_c=2.0$. From Evans and Searles (2002).

where α is the Nosé–Hoover thermostat multiplier. At $t=0$, the field and the thermostat are switched off and the system is allowed to relax to equilibrium. We note that this system is not T-mixing for $t>0$ because states with different energies cannot mix. We can, however, regard it as a canonical sum of independent, non-interacting, T-mixing systems.

Figure 7.1 shows the modulation in the color density of the particles at $t<0$ and the color mixing that occurs as predicted by the Le Chatelier's principle[1] when the field is switched off.

The fluctuation theorem for this system would predict that, although mixing would be the most likely outcome, for small systems and short periods the color modulation could in fact become *stronger.* This demixing violates Le Chatelier's principle. Figure 7.2 shows a histogram of values for the time-integrated dissipation $p(\overline{\Omega}_t t)$ and Figure 7.3 shows that the fluctuation theorem is satisfied for this system.

7.6
Instantaneous Fluctuation Relations (Petersen, Evans, and Williams, 2013)

The covariant dissipation relation, Eq. (3.29), states that $\Omega_\tau(S^{t_1}\mathbf{\Gamma};t_1) = \Omega_{2t_1+\tau}(\mathbf{\Gamma};0)$. This means that

$$\Omega_0(S^{t_1}\mathbf{\Gamma};t_1) = \ln\left(\frac{f\left(S^{t_1}\mathbf{\Gamma};t_1\right)}{f(M^T S^{t_1}\mathbf{\Gamma};t_1)}\right)$$

1) *If a system has a stable equilibrium, then any spontaneous change in its parameters must bring about processes which tend to restore the system to equilibrium* Atkins and De Paula (2006).

$$= \lim_{\delta V_{\Gamma}\to 0} \ln\left(\frac{p\left[\delta V_{\Gamma}\left(S^{t_1}\boldsymbol{\Gamma}\right);t_1\right]}{p[\delta V_{\Gamma}(M^T S^{t_1}\boldsymbol{\Gamma});t_1]}\right)$$

$$= \Omega_{2t_1}(\boldsymbol{\Gamma};0) = \lim_{\delta V_{\Gamma}\to 0} \ln\frac{p[\delta V_{\Gamma}(\boldsymbol{\Gamma});0]}{p[\delta V_{\Gamma}(M^T S^{2t_1+\tau}\boldsymbol{\Gamma});0]} \tag{7.48}$$

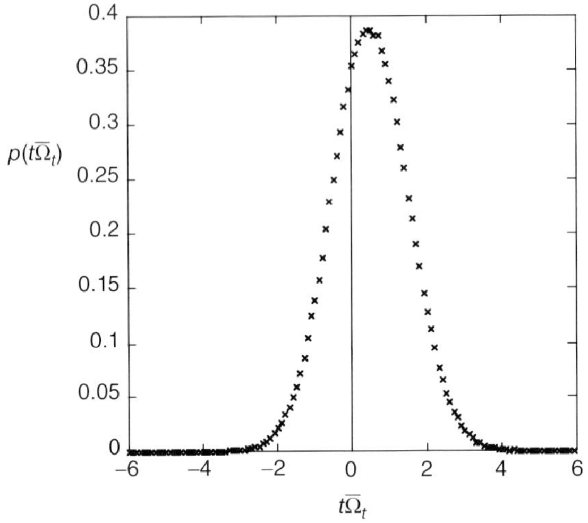

Figure 7.2 Histogram of the distribution of the dissipation function for a system containing a color-separated binary system that is relaxing to equilibrium. Here, $T = 1.0$, $n = 0.4$, $F_c = 2.0$, and $t = 0.4$. From Evans and Searles (2002).

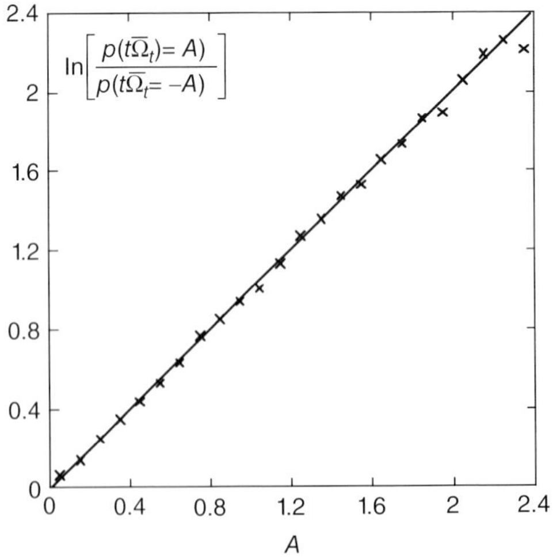

Figure 7.3 Test of the fluctuation theorem given by Eq. (7.46) for a system containing a color-separated binary system that is relaxing to equilibrium. Here, $T = 1.0$, $n = 0.4$, $F_c = 2.0$, and $t = 0.4$. From Evans and Searles (2002).

Thus the ratio of observing trajectories passing through a small phase space volume element near $\delta V_{\Gamma}(S^{t_1}\mathbf{\Gamma})$ at time t_1, compared to the conjugate set of antitrajectories near $\delta V_{\Gamma}(M^T S^{t_1}\mathbf{\Gamma})$ at time t_1, is the same as the probability ratio of observing trajectories near $\delta V_{\Gamma}(\mathbf{\Gamma})$ at time zero, compared to their conjugate antitrajectories at $\delta V_{\Gamma}(M^T S^{2t_1}\mathbf{\Gamma})$ at time zero. This is just the conservation of ensemble members in comoving phase space volume elements.

The functional transient fluctuation formula, Eq. (3.25), can be applied to a trajectories $0, 2t + \tau$, where averaging of the phase variable commences at t:

$$\frac{p(\overline{\phi}_{t,t+\tau} = A)}{p(\overline{\phi}_{t,t+\tau} = -A)} = \frac{\int_{\overline{\phi}_{t,t+\tau}(\mathbf{\Gamma})=A} d\mathbf{\Gamma}\ f(\mathbf{\Gamma};0)}{\int_{\overline{\phi}_{t,t+\tau}(\mathbf{\Gamma})=-A} d\mathbf{\Gamma}^*\ f(\mathbf{\Gamma}^*;0)}$$
$$= \frac{\int_{\overline{\phi}_{t,t+\tau}(\mathbf{\Gamma})=A} d\mathbf{\Gamma}\ f(\mathbf{\Gamma};0)}{\int_{\overline{\phi}_{t,t+\tau}(\mathbf{\Gamma})=A} d\mathbf{\Gamma}\ f(\mathbf{\Gamma};0)e^{-\Omega_{0,2t+\tau}(\mathbf{\Gamma};0)}}$$
$$= \langle e^{-\Omega_{0,2t+\tau}(\mathbf{\Gamma};0)}\rangle^{-1}_{\overline{\phi}_t=A} \tag{7.49}$$

Considering an arbitrary phase function $B(\mathbf{\Gamma})$, which is odd under the time-reversal map M^T, and substituting $\lim_{\tau\to 0}\overline{B}_{t,t+\tau} = B(S^{t_1}\mathbf{\Gamma})$ for the phase variable $\overline{\phi}_{t,t+\tau}$ in the functional transient fluctuation formula, Eq. (7.49), gives

$$\frac{p(B(S^{t_1}\mathbf{\Gamma}) = A)}{p(B(S^{t_1}\mathbf{\Gamma}) = -A)} = \langle \exp[-\Omega_{2t_1}(\mathbf{\Gamma};0)]\rangle^{-1}_{B(S^{t_1}\mathbf{\Gamma})=A}$$
$$= \langle \exp[-\Omega_{2t_1}(\mathbf{\Gamma};0)]\rangle_{B(S^{t_1}\mathbf{\Gamma})=-A} \tag{7.50}$$

Thus the probability ratio of observing opposite values of a phase function at a time t_1 is related to negative exponentials of path integrals of the dissipation from $t = 0$ to $2t_1$ (Petersen, Evans, and Williams, 2013).

7.7
Further Properties of the Dissipation Function (Evans, Searles and Williams, 2008)

Originally, the dissipation function, Eq. (3.4),was defined in order to characterize the ratio of probabilities p of observing infinitesimal sets of phase space trajectories originating at $t=0$ in a phase space volume $\delta V_{\Gamma}(\mathbf{\Gamma})$ to the probability of observing at $t=0$ their time-reversed antitrajectories inside $\delta V_{\Gamma}(M^T S^t\mathbf{\Gamma})$:

$$\frac{p(\delta V_{\Gamma}(\mathbf{\Gamma});0)}{p(\delta V_{\Gamma}(M^T S^t\mathbf{\Gamma});0)} = \exp\left[\int_0^t ds\Omega\,(S^s\mathbf{\Gamma})\right] \tag{7.51}$$

Combining Eq. (7.51) with Eq. (4.3) – which for convenience we repeat here as Eq. (7.52)

$$f(\mathbf{\Gamma};t) = \exp\left[-\int_0^{-t} ds\ \Omega\,(S^s\mathbf{\Gamma})\right] f(\mathbf{\Gamma};0) \tag{7.52}$$

shows that the nonequilibrium N-particle distribution function at time t can be written in terms of the ratio of probabilities of observing conjugate trajectories:

$$f(\mathbf{\Gamma};t) = \frac{p(\delta V_{\mathbf{\Gamma}}(M^T S^{-t}\mathbf{\Gamma});0)}{p(\delta V_{\mathbf{\Gamma}}(\mathbf{\Gamma};0)} f(\mathbf{\Gamma};0) \tag{7.53}$$

We find it remarkable that the measure of *irreversibility* given in Eq. (7.52) by the dissipation function also features so centrally in the dissipation theorem. This shows that this measure of irreversibility is the prime function in determining how a nonequilibrium system will *respond* to a nonequilibrium perturbation or dissipative field.

The dissipation theorem can be used to calculate the ensemble average of an arbitrary phase variable and for arbitrarily strong dissipative fields F_e. In deriving Eq. (7.52), we considered a system that preserves the initial (equilibrium) distribution in the absence of an external dissipative field. Our formalism is sufficiently general to also describe the process of equilibration where there is no external dissipative field, $F_e = 0$, but where dissipation occurs only because the initial distribution is not preserved by the dynamics (i.e., is a nonequilibrium distribution). The results given in Eq. (7.52) apply to this more general circumstance (this includes systems subject to rapid temperature or pressure quenches, and so on, but in which there is no applied *mechanical* dissipative field).

It is interesting to compare a number of different relationships between the distribution function, the dissipation function, and the phase space expansion factor (Evans, Searles, and Williams, 2008). The first such relation is Eq. (7.51). We note that, although the time argument in Eq. (7.52) is negative, the dynamics must still be governed by the field-dependent, thermostatted equations of motion. Rewriting the definition of the time-integrated dissipation (Eq. (3.2)), we have

$$f(S^t\mathbf{\Gamma};0) = \exp\left[-\int_0^t ds\left[\Omega(S^s\mathbf{\Gamma}) + \Lambda(S^s\mathbf{\Gamma})\right]\right] f(\mathbf{\Gamma};0) \tag{7.54}$$

In a nonequilibrium steady state (SS), $\langle\Omega(t)\rangle_{\mathrm{SS}} = -\langle\Lambda(t)\rangle_{\mathrm{SS}}$, which is the phase space compression rate. In the early literature on fluctuation relations, many authors confused the difference between dissipation and phase space compression rate; many thought they were the same quantity not just on average but also instantaneously (as in fluctuation relations). Only in the special case where the initial ensemble is microcanonical and the dynamics is isoenergetic (system of interest *plus* walls and thermostat) is $\Omega(t) = -\Lambda(t),\ \forall t$. In driven thermostatted canonical systems, the instantaneous dissipation is the work (i.e., the change in the energy minus that change due to the heat) divided by $k_{\mathrm{B}}T$, whereas the phase space expansion factor is the heat lost from the system due to the thermostat divided by $k_{\mathrm{B}}T$. As the driving field gets smaller, the absolute difference between the instantaneous values becomes independent of the field.

Lastly, we have the formal solution of the phase continuity equation in its Lagrangian form, Eq. (2.35):

$$f(S^t\mathbf{\Gamma};t) = \exp\left[-\int_0^t ds\Lambda(S^s\mathbf{\Gamma})\right] f(\mathbf{\Gamma};0) \tag{7.55}$$

Comparing Eqs. (7.52), (7.54), and (7.55) clearly shows the difference between dissipation and the phase space compression factor. To this day, these two quantities, namely phase space compression and dissipation, are often confused as being identical.

References

Atkins, P. and De Paula, J. (2006) *Atkin's Physical Chemistry*, Oxford University Press, Oxford.

De Groot, S.R. and Mazur, P. (1984) *Non-Equilibrium Thermodynamics*, Dover Publications, New York.

Evans, D. J., Hoover, W. G., Failor, B. H., Moran, B. & Ladd, A. J. C. (1983). Non-equilibrium molecular-dynamics vis Gauss principle of least constraint. *Phys. Rev. A*, **28**, 1016–1021.

Evans, D.J. and Morriss, G.P. (1990) *Statistical Mechanics of Nonequilibrium Liquids*, Academic Press, London.

Evans, D.J. and Sarman, S. (1993) Equivalence of thermostatted nonlinear responses. *Phys. Rev. E*, **48**, 65–70.

Evans, D.J. and Searles, D.J. (2002) The fluctuation theorem. *Adv. Phys.*, **51**, 1529–1585.

Evans, D.J., Searles, D.J., and Mittag, E. (2001) Fluctuation theorem for hamiltonian systems: Le Chatelier's principle. *Phys. Rev. E*, **63**, 051105.

Evans, D.J., Searles, D.J., and Williams, S.R. (2008) On the fluctuation theorem for the dissipation function and its connection with response theory. *J. Chem. Phys.*, **128**, 014504.

Evans, D.J., Searles, D.J., and Williams, S.R. (2010) On the probability of violations of Fourier's Law for heat flow in small systems observed for short times. *J. Chem. Phys.*, **132**, 024501-1.

Evans, D.J., Williams, S.R., and Rondoni, L. (2012) A mathematical proof of the zeroth "law" of thermodynamics and the nonlinear Fourier "law" for heat flow. *J. Chem. Phys.*, **137**, 194109.

Gallavotti, G. (2008) *The Fermi Pasta Ulam Problem*, Springer, Heidelberg.

Maxwell, J.C. (1872) *Theory of Heat*, Longmans, London.

Maxwell, J.C. (1874) Van Der Waals on the continuity of the gaseous and liquid states. *Nature*, **10**, 477–480.

Mejia-Monasterio, C. and Rondoni, L. (2008) On the fluctuation relation for nosé-hoover boundary thermostatted systems. *J. Stat. Phys.*, **133**, 617.

Petersen, C.F., Evans, D.J., and Williams, S.R. (2013) The instantaneous fluctuation theorem. *J. Chem. Phys.*, **139**, 184106.

Petravic, J. and Harrowell, P. (2005) Linear response theory for thermal conductivity and viscosity in terms of boundary fluctuations. *Phys. Rev. E*, **71**, 061201.

Petravic, J. and Harrowell, P. (2006) Linear response theory for thermal conductivity and viscosity in terms of boundary fluctuations[*Phys. Rev. E* (2005), **71**, 061201], Erratum.. *Phys. Rev. E*, **74**, 049903.

Schmidt, R.K. and Evans, D.J. (1994) Response theory analysis of a thermodynamic temperature quench. *Mol. Phys.*, **83**, 9–17.

Searles, D.J. and Evans, D.J. (2001) A fluctuation theorem for heat flow. *Int. J. Thermophys.*, **22**, 123–134.

Zwanzig, R. (2001) *Nonequilibrium Statistical Mechanics*, Oxford University Press, Oxford.

8
Nonequilibrium Work Relations, the Clausius Inequality, and Equilibrium Thermodynamics

> … suppose our senses sharpened to such a degree that we could trace the motions of molecules as easily as we now trace those of large bodies, … the distinction between work and heat would vanish… .
>
> *(Maxwell 1878, p. 279)*

In previous chapters devoted to the fluctuation, dissipation, and relaxation theorems, once we set the dynamics running at the initial time, or perhaps at time 0^+, at no stage did we change the underlying equilibrium thermodynamic state. (In thermal quenches or pressure quenches, we changed the underlying equilibrium state from $t = 0^-$ to $t = 0^+$.) If at any time we ceased doing work on a driven system, the system would relax back to the (perhaps underlying) thermodynamic equilibrium state we started from at $t = 0^+$. From $t = 0^+$ onward, the equilibrium state for the system with the dissipative field set to zero did not change with time. In equilibrating systems, the system is not initially in equilibrium, but the equilibrium state specified by the zero-field dynamics was unchanging and for T-mixing systems it is unique. In driven systems, the system starts in equilibrium, but is driven out of equilibrium by the dissipative field. If the dissipative field is subsequently set to zero, the system will return toward the *initial* equilibrium state.

As mentioned in Section 2.2, there is no easy mathematical way to tell whether a given field is dissipative or elastic (i.e., it changes the free energy of the underlying equilibrium state). For example, electric fields or strain fields can be either, depending on the temperature and/or density of the system.

The clearest indication that the dissipation is purely dissipative is evidenced in the nonequilibrium partition identity (NPI) (Section 3.3). If the integrated dissipation contained a reversible component ΔW_{rev}, then the left-hand side of the NPI, Eq. (3.12), would have to be equal to $\exp[-\Delta W_{rev}]$ rather than unity.

Another feature of the dynamics studied previously was that the equations of motion, at least after $t = 0^+$, were almost always autonomous. (The exception was that for the ESFT (Evans-Searles fluctuation theorem) we could employ non-autonomous dynamics provided it had a definite parity over a specific time interval, under the time-reversal mapping.) In this chapter we will very frequently

Fundamentals of Classical Statistical Thermodynamics: Dissipation, Relaxation and Fluctuation Theorems, First Edition. Denis J. Evans, Debra J. Searles, and Stephen R. Williams.

discuss non-autonomous systems in which either the Hamiltonian or some thermodynamic state variables change non-autonomously during the dynamics.

Aside
We note that in classical thermodynamics, the free energy differences between two equilibrium states are determined using the work carried out along a quasi-static (i.e., equilibrium, reversible) pathway connecting the two equilibrium states. Of course, in classical thermodynamics one only performs these measurements in the so-called thermodynamic limit where all intensive thermodynamic quantities are independent of the system size and where fluctuations vanish.

Jarzynski (1997a,b) discovered the first of a set of new fluctuation relations that used rapidly traversed nonequilibrium path integrals measured for an ensemble of *nonequilibrium* pathways to provide *equilibrium* thermodynamic information about *small* systems. Although the relationships are formally correct for systems of arbitrary size, in order for these approaches to be useful, the system size *must* be small because the methods rely on fluctuations and the observation of phase space trajectories that are the time-reversed conjugate trajectories to the most probable trajectories (i.e., they are among the *most* improbable trajectories possible within the specified phase space domain). In addition, they are applicable to a small system (such as a single molecule) immersed in a large bath with which it interacts. Although the requirement for small system size may be seen as something of a disadvantage, it turns out to be an *essential* advantage for studying the thermodynamics of small nano- or biosystems, something that classical thermodynamics could not do.

The fundamental reason why the Jarzynski equality (JE) works is because, if we write the nonequilibrium work as a sum of the purely reversible thermodynamic work and the purely irreversible work, the irreversible work satisfies an NPI, leading directly to the JE. We show this in detail in this chapter – see Section 8.6.

The Crooks fluctuation theorem (CFT) (Crooks, 1999) is another such fluctuation relation that is applicable to the same classes of systems as the JE. CFT and the JE were originally developed for determining the difference in free energy of canonical equilibrium states at the same temperature. However, we present a very general formalism for deriving nonequilibrium free energy relations that can be applied in a very wide variety of circumstances.

There is, at the time of writing this book, a huge literature on CFT and JE. When these results were first announced, there appeared in the literature a number of papers claiming CFT and JE were incorrect. By now they are both well-accepted theoretical relations. Although the vast majority of theoretical papers employ stochastic rather than deterministic methods, we will, as always, employ time-reversible deterministic dynamics as first outlined for CFT and JE by ourselves (Evans, 2003). By employing classical deterministic Newtonian dynamics, the only global assumptions we are making are that quantum and relativistic effects are insignificant.

Later in the chapter, we show how to derive the Clausius inequality for a thermal reservoir directly from mechanical considerations (Evans, Williams, and Searles, 2011). In 1854, Clausius proved his inequality by *assuming* the second "Law" of thermodynamics (Clausius, 1854, 1856). In fact, Clausius' statement of the second "Law" is perhaps the most commonly used form of the law. The fact that we can now prove this from the laws of mechanics, completely changes the logical structure of thermodynamics. The "Laws" of thermodynamics, in fact, cease being laws and instead become theorems provable from the time-reversible equations of mechanics and the axiom of causality (Chapter 9). No longer is thermodynamics in apparent contradiction to time-reversible microscopic dynamics. A further consequence of our exposition is that the logical compartmentalization of thermodynamics as being separate from, and indeed in conflict with mechanics and electrodynamics, vanishes. Our microscopic analysis leads directly, in the thermodynamic and quasi-static limits, to the equations of classical thermodynamics.

Our mechanical proof of the Clausius inequality leads to the Clausius inequality for the reservoir, but it leads to different results for the system of interest (soi). It gives meaning to the temperature in the non-quasi-static case where it is a *strict inequality* for the reservoir. As noted by Bertrand, Orr, and Buckingham over a century ago (Bertrand, 1887; Orr, 1904; Buckingham, 1905), in the case of a strict inequality, Clausius gave no logically consistent definition for the temperature. We now say that at any point in a Clausius cycle, the temperature is always the temperature of the underlying equilibrium state to which the system plus reservoir will relax if they are so allowed. A consequence of this is, of course, that the changes in the entropy for the system of interest and the reservoir are always equal and opposite even in the non-quasi-static case. The entropy, in fact, does not "tend to a maximum" as Clausius claimed (Clausius, 1854). It is simply constant, as Gibbs discovered for autonomous Hamiltonian systems. It is dissipation, not entropy, that tends to increase.

8.1
Generalized Crooks Fluctuation Theorem (GCFT) (Evans, Searles, and Williams, 2011)

We consider two closed N-particle systems: 1, 2. These systems may have the same or different Hamiltonians, temperatures, or volumes; it does not matter. Nor does the ensemble matter: microcanonical, canonical, or isothermal isobaric. A protocol, and the corresponding time-dependent dynamics, is then defined that will eventually transform equilibrium system 1 into equilibrium system 2 – at least with respect to taking averages of physical properties. The systems are distinguished by introducing a parameter $\lambda(t)$, which takes on a value λ_1 in system 1 and λ_2 in system 2, and the transformation is also parameterized through $\lambda(t)$ with $\lambda(0) = \lambda_1$ and $\lambda(\tau) = \lambda_2$. The equations of motion are therefore non-autonomous (i.e., they depend *explicitly* on time).

Definitions

We define a *generalized dimensionless "work"* $\Delta X_\tau(\mathbf{\Gamma})$ for a trajectory of duration τ originating from the phase point $\mathbf{\Gamma}$ as

$$\begin{aligned}\exp[\Delta X_\tau(\mathbf{\Gamma})] &\equiv \lim_{\delta V_\Gamma \to 0} \frac{p_{\text{eq},1}(\delta V_\mathbf{\Gamma}(\mathbf{\Gamma};0))Z(\lambda_1)}{p_{\text{eq},2}(\delta V_\mathbf{\Gamma}(S^\tau\mathbf{\Gamma};0)Z(\lambda_2)} \\ &= \frac{f_{\text{eq},1}(\mathbf{\Gamma})d\mathbf{\Gamma}\, Z(\lambda_1)}{f_{\text{eq},2}(S^\tau\mathbf{\Gamma})d(S^\tau\mathbf{\Gamma})Z(\lambda_2)}, \qquad \forall\mathbf{\Gamma} \in D_1 \end{aligned} \tag{8.1}$$

where $Z(\lambda_i)$ is the *partition function* for the system $i = 1{,}2$ and $d\mathbf{\Gamma} = \lim_{\delta V_\Gamma \to 0} \delta V_\mathbf{\Gamma}(\mathbf{\Gamma})$. The *partition function* Z_i is just the normalization factor for the equilibrium distribution function $f_{\text{eq}}(\mathbf{\Gamma})$, and $f_{\text{eq}}(\mathbf{\Gamma}) = \exp[F(\mathbf{\Gamma})]/Z$, where $F(\mathbf{\Gamma})$ is some real single-valued phase function.

Note: one can always multiply the numerator and denominator by a common factor, leaving the equilibrium distribution unchanged. This it related to the fact that in classical statistical thermodynamics the free energies, the entropy, and even the energy are each only defined up to an arbitrary additive constant – see Eq. (5.46).

In Eq. (8.1), D_1 is the accessible phase space domain for system 1, (e.g., coordinates in a fixed special range $(-L, +L)$, and momenta range from $(-\infty, +\infty)$). In Eq. (8.1), $d\mathbf{\Gamma}$ is an infinitesimal phase space volume centered on $\mathbf{\Gamma}$, and $d(S^\tau\mathbf{\Gamma})$ an infinitesimal phase volume centered on $S^\tau\mathbf{\Gamma}$. Without loss of generality, we assume that both equilibrium distribution functions are even functions of the momentum. This implies that we are not moving relative to both systems.

Although the physical significance of the generalized work, X, might seem obscure at this point, we will show that for particular choices of dynamics and ensemble, it is related to important thermodynamic properties and when it is evaluated along quasi-static paths it is in fact a path independent state function.

Before proceeding further with the analysis, it is useful to consider precisely what the generalized work is dependent upon. First, it is a function of the *equilibrium* states 1 and 2. This occurs via the equilibrium distributions appearing in Eq. (8.1) and also the partition functions for those states – see Eq. (8.1). Second, it is a function of the endpoints of the possibly *nonequilibrium* phase space trajectory that takes phase $\mathbf{\Gamma}$ to $S^\tau\mathbf{\Gamma}$. It is also a function of how much heat is gained or lost from the system over the duration of that trajectory. This heat loss determines the ratio of the phase space volumes, $d\mathbf{\Gamma}/d(S^\tau\mathbf{\Gamma})$. Lastly, it is a function of the duration of the trajectories, τ.

The probability of observing ensemble members within the infinitesimal phase volume $d\mathbf{\Gamma}$ centered on the phase vector $\mathbf{\Gamma}$ in the initial equilibrium distribution function $f_{\text{eq},1}(\mathbf{\Gamma})$ is $p_{\text{eq},1}(d\mathbf{\Gamma};0) = f_{\text{eq},1}(\mathbf{\Gamma})d\mathbf{\Gamma}$.

It is very important to note that the time τ is the time at which the parametric change in λ is complete. This means that at time τ this system is *not* in equilibrium: $f(\mathbf{\Gamma};0) = f_{\text{eq},1}(\mathbf{\Gamma})$ but $f(\mathbf{\Gamma};\tau) \neq f_{\text{eq},2}(\mathbf{\Gamma})$ in general. We have seen in Chapter 5 that relaxation to complete thermal equilibrium cannot take place in finite time. The generalized work is defined with respect to two different *equilibrium* distributions and the end points of *finite time* phase space trajectories: $S^s\mathbf{\Gamma}\ :\ \ 0 \le s \le \tau$.

Definition

In order for $\Delta X_\tau(\mathbf{\Gamma})$ to be well defined, $\forall\mathbf{\Gamma} \in D_1$, we must have $S^\tau\mathbf{\Gamma} \in D_2$ and both $f_{\mathrm{eq},1}(\mathbf{\Gamma}) \neq 0$ and $f_{\mathrm{eq},2}(S^\tau\mathbf{\Gamma}) \neq 0$. This is known as the *ergodic consistency for the generalized work*.

We identify $\|\partial S^\tau\mathbf{\Gamma}/\partial\mathbf{\Gamma}\|$ as the Jacobian determinant and note that

$$\left\|\frac{\partial S^\tau\mathbf{\Gamma}}{\partial\mathbf{\Gamma}}\right\| = \frac{d(S^\tau\mathbf{\Gamma})}{d\mathbf{\Gamma}} \tag{8.2}$$

The GCFT considers the probability $p_{\mathrm{eq},f}(\Delta X_t = B \pm dB)$ of observing values of ΔX_t in the range $B \pm dB$ for forward trajectories starting from the initial equilibrium distribution 1, $f_1(\mathbf{\Gamma};0) = f_{\mathrm{eq},1}(\mathbf{\Gamma})$, and the probability $p_{\mathrm{eq},r}(\Delta X_t = -B \mp dB)$ of observing ΔX_t in the range $= -B \pm dB$ for reverse trajectories but starting from the equilibrium distribution given by $f_{\mathrm{eq},2}(\mathbf{\Gamma})$ of system 2.

Consider two equilibrium ensembles from which the initial trajectories can be selected with known equilibrium distributions: $f_{\mathrm{eq},1}(\mathbf{\Gamma})$ and $f_{\mathrm{eq},2}(\mathbf{\Gamma})$.

If initially we select phases from $f_{\mathrm{eq},1}(\mathbf{\Gamma})$ and employ a particular protocol (labeled "f") and the corresponding time-dependent dynamics defined by a parameter $\lambda_f(s)$ with $\lambda_f(0) = \lambda_1$ and $\lambda_f(\tau) = \lambda_2$, then the probability that the phase variable defined in Eq. (8.1) takes on the value B is given by

$$p_{\mathrm{eq},1}(\Delta X_{\tau,f} = B \pm dB) = \int_{\Delta X_{\tau,f}=B\pm dB} d\mathbf{\Gamma}\ f_{\mathrm{eq},1}(\mathbf{\Gamma}) \tag{8.3}$$

If initially we select phases from $f_{\mathrm{eq},2}(\mathbf{\Gamma})$ with a particular protocol (labeled "r") which is the time-reverse of (f), $\lambda_r(s) = \lambda_f(\tau - s)$, and corresponding time-dependent dynamics so that $\lambda_r(0) = \lambda_2$ and $\lambda_r(\tau) = \lambda_1$, then the probability that the phase variable defined in Eq. (8.1) takes on the value $-B$ is given by

$$p_{\mathrm{eq},2}(\Delta X_{\tau,f} = -B \mp dB) = \int_{\Delta X_{\tau,f}=-B\mp dB} d\mathbf{\Gamma} f_{\mathrm{eq},2}(\mathbf{\Gamma}).$$

We note that a trajectory starting at point $\mathbf{\Gamma}$ and evolved forward in time with the forward protocol to the point $S^\tau\mathbf{\Gamma}$ will be related by a time-reversal mapping to a trajectory starting at $M^T S^\tau\mathbf{\Gamma}$ and evolving with the time-reverse protocol. If $S^\tau_{f/r}$ is the time evolution operator with forward/reverse protocol,

$$M^T S^\tau_r M^T S^\tau_f \mathbf{\Gamma} = \mathbf{\Gamma} \tag{8.4}$$

Now we look at the ratio of these two probabilities:

$$\frac{p_{\mathrm{eq},1}(\Delta X_{\tau,f} = B \pm dB)}{p_{\mathrm{eq},2}(\Delta X_{\tau,r} = -B \mp dB)} = \frac{\displaystyle\int_{\Delta X_{\tau,f}(\mathbf{\Gamma})=B\pm dB} d\mathbf{\Gamma} f_{\mathrm{eq},1}(\mathbf{\Gamma})}{\displaystyle\int_{\Delta X_{\tau,r}(\mathbf{\Gamma})=-B\mp dB} d\mathbf{\Gamma} f_{\mathrm{eq},2}(\mathbf{\Gamma})} \tag{8.5}$$

If $\Delta X_{\tau,f}(\mathbf{\Gamma}) = B$, then from Eq. (8.1) we know that $\Delta X_{\tau,r}(M^T S^\tau \mathbf{\Gamma}) = -B$. We can therefore write Eq. (8.5) as

$$\begin{aligned}
\frac{p_{\text{eq},1}(\Delta X_{\tau,f} = B \pm dB)}{p_{\text{eq},2}(\Delta X_{\tau,r} = -B \mp dB)} &= \frac{\displaystyle\int_{\Delta X_{\tau,f}(\mathbf{\Gamma})=B\pm dB} d\mathbf{\Gamma} f_{\text{eq},1}(\mathbf{\Gamma})}{\displaystyle\int_{\Delta X_{\tau,r}(\mathbf{\Gamma})=-B\mp dB} d\mathbf{\Gamma} f_{\text{eq},2}(\mathbf{\Gamma})} \\
&= \frac{\displaystyle\int_{\Delta X_{\tau,f}(\mathbf{\Gamma})=B\pm dB} d\mathbf{\Gamma} f_{\text{eq},1}(\mathbf{\Gamma})}{\displaystyle\int_{\Delta X_{\tau,r}(M^T S^\tau \mathbf{\Gamma})=-B\mp dB} d(M^T S^\tau \mathbf{\Gamma}) f_{\text{eq},2}(M^T S^\tau \mathbf{\Gamma})} \\
&= \frac{\displaystyle\int_{\Delta X_{\tau,f}(\mathbf{\Gamma})=B\pm dB} d\mathbf{\Gamma} f_{\text{eq},1}(\mathbf{\Gamma})}{\displaystyle\int_{\Delta X_{\tau,f}(\mathbf{\Gamma})=B\pm dB} d\mathbf{\Gamma} \exp[-\Delta X_{\tau,f}(\mathbf{\Gamma})] f_{\text{eq},1}(\mathbf{\Gamma}) Z(\lambda_1)/Z(\lambda_2)} \\
&= \exp[B] \frac{Z(\lambda_2)}{Z(\lambda_1)}
\end{aligned} \tag{8.6}$$

To obtain the third line, we have used Eq. (8.1), noting that the Jacobian is unchanged on time-reversal mapping, $d(M^T S^\tau \mathbf{\Gamma})/d(S^\tau \mathbf{\Gamma}) = 1$, and that the value of the equilibrium distribution is unchanged on time-reversal mapping of the phase point.

Definition

Equation (8.6) is the *generalized Crooks fluctuation relation* (GCFR) and its derivation is the GCFT.

In the derivation of Eq. (8.6), we have not placed restrictions on the dynamics, apart from the conditions of ergodic consistency and reversibility (Eq. (8.4)). Therefore, the nonequilibrium system could evolve under homogeneously thermostatted dynamics, or with thermostats far from the system of interest, or even with no thermostat. The details of the thermostatting will not change the validity of Eq. (8.6).

However, even if the system is thermostatted, as previously mentioned, at time τ the system that is evolving from $f_{\text{eq},1}(\mathbf{\Gamma})$ will not have relaxed *to* the equilibrium distribution $f_{\text{eq},2}(\mathbf{\Gamma})$ (or vice versa). In fact, complete relaxation never takes place in finite time – see Sections 5.3 and 5.4. We can compute the change in the generalized work going from a time τ to time $\tau + s$. During this interval, there is no parametric change and the system simply relaxes toward the equilibrium state 2 produced by the thermostatted equations of motion. From Eq. (8.1) we see that

$$\begin{aligned}
\exp[\Delta X_{\tau+s}(\mathbf{\Gamma}) - \Delta X_\tau(\mathbf{\Gamma})] &= \frac{f_{\text{eq},1}(\mathbf{\Gamma}) d\mathbf{\Gamma}\ Z(\lambda_1)}{f_{\text{eq},2}(S^{\tau+s}\mathbf{\Gamma}) d(S^{\tau+s}\mathbf{\Gamma}) Z(\lambda_2)} \frac{f_{\text{eq},2}(S^\tau \mathbf{\Gamma}) d(S^\tau \mathbf{\Gamma}) Z(\lambda_2)}{f_{\text{eq},1}(\mathbf{\Gamma}) d\mathbf{\Gamma}\ Z(\lambda_1)} \\
&= \frac{f_{\text{eq},2}(S^\tau \mathbf{\Gamma}) d(S^\tau \mathbf{\Gamma})}{f_{\text{eq},2}(S^{\tau+s}\mathbf{\Gamma}) d(S^{\tau+s}\mathbf{\Gamma})}, \quad \forall s > 0
\end{aligned} \tag{8.7}$$

If we look at the second line of Eq. (8.7), we recognize that it is simply the exponential of the integrated dissipation function (3.2), $\Omega_{\text{eq},2,s}(S^\tau\boldsymbol{\Gamma})$, defined in the ESFT for *equilibrium* system 2 evaluated at a phase $S^\tau\boldsymbol{\Gamma}$ and integrated for a time s. It is important to note that both the numerator and the denominator of Eq. (8.7) involve *forward* time integrations from system 2 equilibrium (i.e., there is no forward and reverse as in Eq. (8.1)). Therefore

$$\begin{aligned}[\Delta X_{\tau+s}(\boldsymbol{\Gamma}) - \Delta X_\tau(\boldsymbol{\Gamma})] &\equiv \Delta X_s(S^\tau\boldsymbol{\Gamma}) \\ &= \ln \frac{f_{\text{eq},2}(S^\tau\boldsymbol{\Gamma})}{f_{\text{eq},2}(S^{\tau+s}\boldsymbol{\Gamma})} - \int_0^\tau ds\, \Lambda(S^{\tau+s}\boldsymbol{\Gamma}) \\ &= \Omega_{\text{eq},2,s}(S^\tau\boldsymbol{\Gamma}) = 0, \quad \forall\boldsymbol{\Gamma} \in D_2, \quad \forall s > 0 \end{aligned} \tag{8.8}$$

The last line is identically zero because the dissipation function $\Omega_{\text{eq}}(\boldsymbol{\Gamma})$ for all equilibrium systems is identically zero – see Section 3.7 – and we know from the equilibrium relaxation theorem that the system does eventually relax toward the unique, ergodic, dissipationless equilibrium state of system 2!

Equation (8.8) means that we can fix the parametric change interval at τ and take the limit $\lim_{s\to\infty} \Delta X_{\tau+s}(S^{\tau+s}\boldsymbol{\Gamma}) = \Delta X_\tau(S^\tau\boldsymbol{\Gamma})$ without affecting Eq. (8.6). Thus we can allow the end point distribution of states $\lim_{s\to\infty} f(S^{\tau+s}\boldsymbol{\Gamma}; \tau + s) \to f_{\text{eq},2}(S^{\tau+s}\boldsymbol{\Gamma})$ to relax arbitrarily close to equilibrium, thereby generating the "equilibrium" distribution of the states required for the ensemble of reverse trajectories. We can then generate the reverse trajectories by reversing the entire protocol from the phase vectors $\{S^{\tau+s}(\boldsymbol{\Gamma})\}$.

In order for this process to work, the transient states that start at time τ must be T-mixing, or the final equilibrium distribution $f_{\text{eq},2}(\boldsymbol{\Gamma})$ must be preserved by the dynamics *and* that equilibrium distribution must be mixing. The first condition is true if and only if the second set of conditions is true. If these two sets of conditions are not met, then some Monte Carlo process needs to be used to generate the second equilibrium distribution of states. In this case, there will, of course, be difficulties in *defining* the equilibrium states we are trying to study.

8.2 Generalized Jarzynski Equality (GJE)

The generalized Jarzynski equality (GJE) (Jarzynski, 1997a) can be thought of as the analog of the NPI evaluated for the generalized work. We say "analog" because the introduction of forward and reverse paths in the definition of the generalized work is quite different from the use of forward-only paths for the dissipation function.

The derivation of the GJE from the GCFT is trivial. The simplest approach is to obtain GJE by integration of the GCFT, Eq. (8.6):

$$\begin{aligned}\langle \exp[-\Delta X_\tau(\mathbf{\Gamma})]\rangle_{\mathrm{eq},1} &= \int_{-\infty}^{+\infty} dB\, p_f(\Delta X_\tau = B)\exp(-B) \\ &= \int_{-\infty}^{+\infty} dB\, p_r(\Delta X_\tau = -B)\frac{Z(\lambda_2)}{Z(\lambda_1)} \\ &= \frac{Z(\lambda_2)}{Z(\lambda_1)}\end{aligned} \tag{8.9}$$

If the two states had the same partition functions, (8.9) would, *superficially*, look almost identical to the NPI. The proof is line by line analogous to that given in Section 3.3 for the NPI. We also note that with the change of variables, the domain of integration may change.

As is the case for NPI, GJE can also be derived straightforwardly from Eq. (8.1). Here we compute the relevant average directly:

$$\begin{aligned}\langle \exp[-\Delta X_\tau(\mathbf{\Gamma})]\rangle_{\mathrm{eq},1} &= \int_{D_1} d\mathbf{\Gamma}\; f_{\mathrm{eq},1}(\mathbf{\Gamma})\frac{f_{\mathrm{eq},2}(S^\tau\mathbf{\Gamma})Z(\lambda_2)d(S^\tau\mathbf{\Gamma})}{f_{\mathrm{eq},1}(\mathbf{\Gamma})Z(\lambda_1)d\mathbf{\Gamma}} \\ &= \int_{D_2} d(S^\tau\mathbf{\Gamma})\; f_{\mathrm{eq},2}(S^\tau\mathbf{\Gamma})\frac{Z(\lambda_2)}{Z(\lambda_1)} \\ &= \frac{Z(\lambda_2)}{Z(\lambda_1)}\end{aligned} \tag{8.10}$$

where the brackets $\langle\cdots\rangle_{\mathrm{eq},1}$ denote an equilibrium ensemble average over the initial equilibrium distribution. We also note that with the change of variables the domain of integration may change. There is obviously a trivial $\langle \exp[-\Delta X]\rangle_{\mathrm{eq},2}$ analog to Eqs. (8.9) and (8.10).

The validity of Eqs. (8.9) and (8.10) requires that $\forall\mathbf{\Gamma} \in D_1,\; f_{\mathrm{eq},1}(\mathbf{\Gamma}) \neq 0$, which implies $\forall S^\tau\mathbf{\Gamma} \in D_2,\; f_{\mathrm{eq},2}(S^\tau\mathbf{\Gamma}) \neq 0$. This is the *ergodic consistency condition for the generalized work.* As mentioned previously, one can talk of the ergodic consistency for the theoretical equilibrium distribution as well as the ergodic consistency of the empirical data that one has at one's disposal. Even if the theoretical distributions are ergodically consistent, the observed data may not be. This will lead to biased or skewed estimates for the partition function ratio, as was discussed previously in detail.

From the first line of Eq. (8.9), it is clear that trajectories for which the value of ΔX_τ is negative have a contribution to the ensemble average, which is exponentially enhanced. Therefore, in order to obtain numerical convergence of the ensemble average, it is important that these trajectories are sufficiently well sampled. Many recent studies have addressed this issue and have developed algorithms to improve convergence (Adjanor, Athènes, and Calvo, 2006; Kofke, 2006; Wu and Kofke, 2005a–c; Macfadyen and Andricioaei, 2005; Lechner and Dellago, 2007; Lua and Grosberg, 2005; Shirts and Pande, 2005; Ytreberg, Swendsen, and Zuckerman, 2006; Lechner *et al.*, 2006; Schmiedl and Seifert, 2007; Vaikuntanathan and Jarzynski, 2008). If the averaging process is not sufficiently exhaustive for these possibly extremely rare events to be observed, Eqs. (8.9) and (8.10) will give incorrect results and the statistics for the average will be highly skewed.

This skewness makes it extremely hard to estimate the statistical uncertainties for the average of the left-hand side of Eqs. (8.9) and (8.10).

We now outline a simple but effective way to improve the estimate for the partition function ratio. As noted in Section 3.3, numerical evaluation of the NPI can be improved by guaranteeing ergodic consistency of the numerical data, and a similar approach can be used to treat the partition function ratio. We can define ranges $[\ln(Z(\lambda_1)/Z(\lambda_2)) - b, \ln(Z(\lambda_1)/Z(\lambda_2)) + b] \equiv [B_1, B_2]$ for the forward protocol and $[\ln(Z(\lambda_2)/Z(\lambda_1)) - b, \ln(Z(\lambda_2)/Z(\lambda_1)) + b] \equiv [-B_2, -B_1]$ for the reverse protocol within which the *observed* distributions of the generalized work will be nonzero. Then the normalized probability distributions for the dimensionless generalized work over the restricted range can be computed as

$$\hat{p}_f(\Delta X_\tau) = p_f(\Delta X_\tau) / \int_{B_1}^{B_2} dA \ p_f(\Delta X_\tau = A) = p_f(\Delta X_\tau)/b_f$$

$$\hat{p}_r(\Delta X_\tau) = p_r(\Delta X_\tau) / \int_{-B_2}^{-B_1} dA \, p_r(\Delta X_\tau = A) = p_r(\Delta X_\tau)/b_r \tag{8.11}$$

Note that over the restricted range

$$\begin{aligned} b_f \hat{p}_f(\Delta X_\tau) &= p_f(\Delta X_\tau), \quad \forall (B_1 < \Delta X_\tau < B_2) \\ b_r \hat{p}_r(\Delta X_\tau) &= p_r(\Delta X_\tau), \quad \forall (-B_2 < \Delta X_\tau < -B_1) \end{aligned} \tag{8.12}$$

where the factors b_f, b_r are both less than unity because of the range restriction for $\hat{p}_{f/r}(\Delta X)$ and the fact that all four probability distributions are normalized over their respective three different ranges. Now compute the following:

$$\begin{aligned} \frac{b_f}{b_r} \langle \exp[-\Delta X_\tau(\boldsymbol{\Gamma})] \rangle_{\mathrm{eq},1,B_1 < \Delta X_\tau < B_2} &= \frac{b_f}{b_r} \int_{B_1}^{B_2} dA \ \hat{p}_f(\Delta X_\tau = A) \exp(-A) \\ &= \frac{1}{b_r} \int_{B_1}^{B_2} dA \ p_f(\Delta X_\tau = A) \exp(-A) \\ &= \frac{1}{b_r} \int_{B_1}^{B_2} dA \ p_r(\Delta X_\tau = -A) \frac{Z(\lambda_2)}{Z(\lambda_1)} \\ &= \int_{-B_2}^{-B_1} dA \ \hat{p}_r(\Delta X_\tau = A) \frac{Z(\lambda_2)}{Z(\lambda_1)} \\ &= \frac{Z(\lambda_2)}{Z(\lambda_1)} \end{aligned} \tag{8.13}$$

This restricted range estimate for the partition function ratio will have a substantially reduced skewness. This is because the *numerical data* satisfies ergodic consistency. In the unrestricted range case, one can easily have, for the probabilities inferred from numerical data, that $p_f(\Delta X) = 0$ while $p_r(-\Delta X) \neq 0$, or vice versa. This means that in *almost all* experiments, the average for the left-hand side of Eqs. (8.9) and (8.10) approaches its mean value from either above or below. Ever more sampling may sample the possibly extreme events that move the average to below or above the mean, respectively. The problem is that these rare events have a large effect on the mean. In the restricted range case this cannot happen. The range for the work has been restricted so that these conjugate sets of trajectories and antitrajectories are actually observed in the experimental data.

These problems concerning ergodic consistency are more pronounced for Jarzynski relations than for the Crooks relation because in the latter we see immediately if this condition is broken. The problem is hidden in the Jarzynski relation because, if the theoretical range for the work is unbounded, the generalized Jarzynski relation, Eqs. (8.9) and (8.10), is an infinite range relation and for actual experimental data ergodic consistency *will* always break down eventually.

This observation concerning ergodic consistency in the numerical data also has an immediate impact on the calculation of free energy differences as the thermodynamic limit is approached. For sufficiently large systems, the bounds in Eq. (8.13) will be zero! In this case, the Crooks and Jarzynski relations become impossible to use. In this case, the averages *must* be calculated in finite systems for a series of system sizes and then extrapolated to obtain the value in the thermodynamic limit. This may not be an easy or cheap task. If you apply GCFT or GJE to macroscopic systems, you will *never* observe the required fluctuations and the GJE can only be applied in the infinitely slow quasi-static (qs) limit:

$$\begin{aligned}\lim_{\tau\to\infty}\lim_{N\to\infty}\left\langle \exp\left[-\Delta X_\tau(\mathbf{\Gamma})\right]\right\rangle_{\mathrm{eq},1} &= \lim_{\tau\to\infty}\lim_{N\to\infty}\exp\left[\left\langle -\Delta X_\tau(\mathbf{\Gamma})]\right\rangle_{\mathrm{qs,eq},1}\right] \\ &= \frac{Z(\lambda_2)}{Z(\lambda_1)}\end{aligned} \tag{8.14}$$

The generalized Jarzynski and Crooks relations *cannot* be applied to macroscopic systems using nonequilibrium pathways.

Equations (8.9), 8.11, and (8.13) are very general, and they even apply to stochastic dynamics. Obviously, the paths do not need to be quasi-static paths as in traditional thermodynamics. These equations are independent of the particular protocol, provided ergodic consistency holds. In fact, it is possible to average over the initial ensemble *and* an ensemble of protocols since the final answer is protocol- or path-independent.

As is the case for the NPI, the GJE can be, as we have seen, proved from the GCFR. However, the reverse is not true because the fluctuation relations contain more information than either the NPI or GJE.

8.3
Minimum Average Generalized Work

We now derive a further simple corollary of the GJE. From Eq. (8.10), we see that

$$\begin{aligned}\frac{Z(\lambda_2)}{Z(\lambda_1)} &= \langle \exp[-\Delta X_\tau]\rangle_1 \\ &= \exp[-\langle \Delta X_\tau\rangle_1]\langle \exp[-\Delta X_\tau + \langle \Delta X_\tau\rangle_1]\rangle \\ &\geq \exp[-\langle \Delta X_\tau\rangle_1]\langle 1 - \Delta X_\tau + \langle \Delta X_\tau\rangle_1\rangle \\ &= \exp[-\langle \Delta X_\tau\rangle_1]\end{aligned} \tag{8.15}$$

In deriving this relation, we have used the fact that $e^x \geq 1 + x, \;\; \forall x \in \mathbb{R}$. Taking the logarithms of both sides and then multiplying both sides by -1 we get

$$\langle \Delta X_\tau \rangle \geq \ln \left[\frac{Z(\lambda_1)}{Z(\lambda_2)} \right] \tag{8.16}$$

This is clearly the analog of the second law inequality for systems of changing free energy. Some authors refer to work inequalities like Eq. (8.16) as the Clausius inequality; however, we reserve that term for *cyclic* inequalities of the *heat*, since as Planck remarked (Planck, 1945), "this is the form of the Second Law first enunciated by Clausius."

In actual systems, the right-hand side will turn out to be a dimensionless free energy difference. For example, if systems 1 and 2 are canonical and at the same temperature and have the same number of particles and volume, as we will see later (Section 8.7), $\ln[Z_1/Z_2] = \beta\Delta A_{21} = \beta(A_2 - A_1)$ and $\Delta X = \beta \int_0^\tau ds \;\; W(s)$, where W denotes the work (i.e., the change internal energy minus that change caused by the heat), and A_i is the Helmholtz free energy of system i. The minimum average work inequality implies in this case $\Delta W_{21} \geq \Delta A_{21}$. The minimum work is expended if the path is reversible or quasi-static, in which case that work is, in fact, the difference in the Helmholtz free energies divided by $k_B T$.

If the parametric protocol takes us around a closed cycle that is defined in terms of the parameter $\lambda(t)$, we see that since by definition $Z_1/Z_2 = 1$

$$\oint ds \langle \dot{X}(s) \rangle = \oint \langle dX \rangle \geq 0 \tag{8.17}$$

The ensemble average of the *cyclic integral of the generalized work is nonnegative.*

Definition

We call Eq. (8.17) the *nonequilibrium cyclic inequality for the generalized work.*

Equation (8.17) for the generalized work is very different from the corresponding cyclic integral of the heat. For the work, we simply execute the protocol cycle. For the heat, as we will soon see, we have to complete the cycle many times and wait until the system settles into a periodic response to the cyclic protocol before we can apply the cyclic integral of the heat. Not all systems do settle into a cyclic response. In these cases, we can say nothing about the cyclic integral of the heat.

The reason for the difference is that, if we execute the cycle once along a nonequilibrium path, when the parameter reaches the value λ_2, even though the system is not yet in equilibrium there is, by Eq. (8.8), no further change in the work during the long relaxation process. The heat does change during this relaxation, however!

We first introduced the definition of quasi-static processes in Section 5.7. To make this chapter easier to read, we repeat this definition here.

Definition

If the dynamics is macroscopically reversible, the cyclic integral can be zero only if the cycle is *thermodynamically* reversible or *quasi-static*. A pathway is traversed quasi-statically if the average work for a forward path is equal and opposite the average work for the reversed path.

The cyclic integral of the generalized work for a quasi-static cycle is zero. The proof is obvious. The fact that the cyclic integral of the generalized work is zero, that is, $\oint dX = 0$, also implies that

$$_{\text{qs}}\int_i^f dX = \text{independent of path} \tag{8.18}$$

where the subscript "qs" denotes the fact that the integral is for a quasi-static or thermodynamically reversible pathway. The proof of Eq. (8.18) is obvious. Construct a reversible cycle $i \to f, f \to i$. The cyclic integral must be zero, so if we vary the pathway for the return leg $f \to i$, we must always get the same value for the integrated reversible work, independently of the precise path.

Definition

Finally, we can see that if the integral of the generalized work for paths is independent of the pathway, then that integral must be a *state function* (i.e., a function only of the initial and final states of the system). In fact, this is why the seemingly abstract generalized work defined in Eq. (8.1) is so important. The generalized work for a thermodynamically reversible, or quasi-static, pathway is always a history- and path-independent function of the thermodynamic states of the system at the end points of the path.

Example

What we have proved above is true if quasi-static paths exist, but we have not shown they do. Consider the case where the parameter $\lambda(t)$ is equal to the strain, and suppose we wish to strain the crystal of volume V though an angle $\delta\gamma$. For simplicity, suppose that our protocol is to strain the crystal at a constant rate $\dot{\gamma}$. The time taken to increase the strain from zero to $\delta\gamma$ is $\delta\gamma/\dot{\gamma}$. For small strains, we expect that

$$\lim_{\dot{\gamma},\gamma\to 0} \langle P_{xy}(t)\rangle = -G_0\gamma(t) - \eta_{0^+}\dot{\gamma}, \quad 0 < t < {}^{\delta\gamma}/_{\dot{\gamma}} \tag{8.19}$$

where G_0 is the zero-frequency elastic shear modulus, which will in the end give the quasi-static or macroscopically reversible work, and η_{0+} is the limiting zero-frequency shear viscosity of the crystal. (Note: You cannot speak of *the* zero-frequency shear viscosity of a solid!) The generalized work turns out to be

$$\begin{aligned}
\lim_{\dot{\gamma}\to 0}\langle \Delta X\rangle &= -\lim_{\dot{\gamma}\to 0}\beta V\int_0^{\delta\gamma/\dot{\gamma}} ds\langle P_{xy}(s)\rangle\dot{\gamma} \\
&= \lim_{\dot{\gamma}\to 0}\beta V\int_0^{\delta\gamma/\dot{\gamma}} ds\,[G_0\dot{\gamma}^2 s + \eta_{0+}\dot{\gamma}^2] \\
&= \lim_{\dot{\gamma}\to 0}\beta V\left[\frac{G_0\dot{\gamma}^2\delta\gamma^2}{2\dot{\gamma}^2} + \frac{\eta_{0+}\dot{\gamma}^2\delta\gamma}{\dot{\gamma}}\right] \\
&= \lim_{\dot{\gamma}\to 0}\beta V\left[\frac{G_0\delta\gamma^2}{2} + \eta_{0+}\dot{\gamma}\delta\gamma\right] \\
&= \frac{\beta V G_0\delta\gamma^2}{2}
\end{aligned} \tag{8.20}$$

The second viscous term is always positive independent of the sign of $\delta\gamma$, $\dot{\gamma}$, whereas the first term is reversible. It can be positive or negative. Obviously, if we strain the crystal though a cycle, the first term will then also vanish while for finite strain rates the second term will satisfy the inequality Eq. (8.17).

8.4 Nonequilibrium Work Relations for Cyclic Thermal Processes (Williams, Searles, and Evans, 2008)

We wish to consider a realistic model of a system that is driven away from equilibrium by a reservoir whose temperature is changing. For this case, the simple parametric change in the Hamiltonian or external field usually employed in the derivation of the GJE or the GCFT is not applicable, and care is needed in developing the physical mechanisms.

Here we could address this issue by considering an system of interest containing some very slowly relaxing constituents, such as soft matter or pitch, in contact with a rapidly relaxing reservoir. The reservoir may be formed from a copper block or another highly thermally conductive material. Changing the temperature of the reservoir (say with a thermostatically controlled heat exchanger) then drives the system of interest out of equilibrium. The change in temperature is slow enough so that the reservoir may be treated, to high accuracy, as undergoing a quasi-static temperature change. The slowly relaxing system of interest is far from equilibrium. We employ the GCFT and the GJE to describe this system. Importantly, the quantities that appear in the theory are physically measurable variables.

Another mechanism for achieving the required result would be (following Planck (1945)) to have an ensemble of large equilibrium thermostats that can be thermally coupled to the system of interest in a protocol sequence. If these thermostats are large, they can be regarded as being in thermal equilibrium. If they are sufficiently remote from the system of interest, there is no way the system of interest can "know" the precise mathematical details of how heat is ultimately taken from or added to the system of interest.

For convenience from a theoretical perspective, we choose the Nosé–Hoover thermostatting mechanism, and the equations of motion, including the thermostat multiplier, are then

$$\begin{aligned}
\dot{\mathbf{q}}_i &= \frac{\mathbf{p}_i}{m} \\
\dot{\mathbf{p}}_i &= \mathbf{F}_i(\mathbf{q}) - S_i(\alpha(\mathbf{\Gamma})\mathbf{p}_i + \mathbf{F}_{\text{th}}) \\
\dot{\alpha} &= \left(\frac{\sum_{i=1}^{N} S_i \mathbf{p}_i \cdot \mathbf{p}_i / m}{3\left(N_{\text{th}} - 1\right) k_B T(t)} - 1 \right) \frac{1}{\tau_\alpha^2}
\end{aligned} \tag{8.21}$$

where τ_α is an arbitrary Nosé–Hoover time constant. The value of $T(t)$ is the target temperature of the thermostat (i.e., the temperature of the underlying equilibrium state at any time t, during the execution of the protocol), and $S_i = 0, 1$ is a switch that controls which particles are coupled to the Nosé–Hoover thermostat – $\sum_{i=1}^{N} S_i = N_{\text{th}}$. In our model, the particles that are coupled to the thermostat can be taken to be remote from the system of interest. This ensures that the particles in the system of interest are ignorant of the precise details of this unphysical thermostat. These thermostatted particles are also subject to a fluctuating force $\mathbf{F}_{\text{th}}$, which is chosen to ensure that the total momentum of the thermostatted particles is identically zero, $\sum_{i=1}^{N} S_i \mathbf{p}_i \equiv \mathbf{p}_{\text{th}} = \mathbf{0}$.

The extended, time-dependent internal energy is

$$H_E(\mathbf{\Gamma}, \alpha, t) = H_0(\mathbf{\Gamma}) + \frac{3(N_{\text{th}} - 1)}{2} k_B T(t) \alpha^2 \tau_\alpha^2 \tag{8.22}$$

and the extended phase space of the system is $\mathbf{\Gamma}' = (\mathbf{\Gamma}, \alpha)$. The phase continuity equation, Eq. (2.34), states $df/dt = -\Lambda f$, and using Eq. (8.21) it is possible to show that

$$\begin{aligned}
k_B T(t) \Lambda(t) &= k_B T(t) \left(\frac{\partial}{\partial \mathbf{\Gamma}} \cdot \dot{\mathbf{\Gamma}}(t) + \frac{\partial}{\partial \alpha} \dot{\alpha}(t) \right) \\
&= -3(N_{\text{th}} - 1) k_B T(t) \alpha(t) = -\dot{Q}_{\text{th}}(t)
\end{aligned} \tag{8.23}$$

where $\dot{Q}_{\text{th}}$ is the rate of decrease in H_E due to the thermostat or, equivalently, the rate of increase of energy in the external thermostat. From the relaxation theorem (Chapter 5), the unique equilibrium distribution function for this system at a fixed temperature T is then

$$\begin{aligned}
f_{\text{eq}}(\mathbf{\Gamma}; T, \alpha) &= \frac{\tau_\alpha \sqrt{3(N_{\text{th}} - 1)/(2\pi)}}{Z_c(T)} \exp(-\beta H_E(\mathbf{\Gamma}, T, \alpha)) \delta(\mathbf{p}_{\text{th}}) \\
&= \frac{\exp(-\beta H_E(\mathbf{\Gamma}, T, \alpha)) \delta(\mathbf{p}_{\text{th}})}{Z_{c,E}(T)}
\end{aligned} \tag{8.24}$$

where $Z_c(T)$ is the canonical partition function for the system of interest, and $Z_{c,E}(T)$ is the corresponding partition function for the extended system; $\lambda(t) \equiv T(t)$.

We now consider applying the GCFR, Eq. (8.6), when a thermal rather than a mechanical process occurs. Consider a thermostatted system of N particles whose target temperature is changed from T_1 to T_2 over a period $0 < t < \tau$. We do not change the Hamiltonian during this process. For simplicity, we consider a canonical ensemble for the two equilibrium states, Eq. (8.22), and use the equations of motion (8.21). The temperature dependence of the reservoir is achieved by making the Nosé–Hoover target temperature $T(t)$ in Eq. (8.21) a time-dependent parameter.

From Eqs. (8.1) and (8.21)–(8.23), we see that the generalized dimensionless work in this extended canonical system is

$$\Delta X_{E,\tau}(\mathbf{\Gamma}'; 0, \tau) = \beta_2 H_E(S^\tau \mathbf{\Gamma}') - \beta_1 H_E(\mathbf{\Gamma}') + \int_0^\tau dt\ \beta(t)\dot{Q}_{\text{th}}(S^t \mathbf{\Gamma}') \tag{8.25}$$

where $\beta(t) = 1/(k_{\text{B}} T(t))$ is the inverse time-dependent target temperature. One can immediately see that this is a form of work for the extended system because it gives the extended system energy change minus the heat increase in the system of interest: $dW_E = dU_E + dQ_{\text{th}} = dU_E - dQ_{\text{soi}}$. Following Planck, dQ_{th} is the heat transferred *to* the thermal reservoir (th) and dQ_{soi} is the heat transferred to the system of interest, which will be equal and opposite. Now if we take the derivative of the extended Hamiltonian while the temperature is changing, but with no other external agent acting on the system, we obtain using Eqs. (8.21) and (8.22)

$$\frac{d}{dt} H_E(S^t \mathbf{\Gamma}') = -\dot{Q}_{\text{th}}(S^t \mathbf{\Gamma}') + \frac{3}{2}(N_{\text{th}} - 1) k_{\text{B}} \dot{T}(t) \alpha^2(t) \tau_\alpha^2 \tag{8.26}$$

We then obtain

$$\frac{d}{dt}[\beta(t) H_E(S^t \mathbf{\Gamma}')] = -\beta(t) \left[H_0\left(S^t \mathbf{\Gamma}'\right) \frac{\dot{T}(t)}{T(t)} + \dot{Q}_{\text{th}}(S^t \mathbf{\Gamma}') \right] \tag{8.27}$$

and combining Eqs. (8.27) and (8.25), the generalized "power" for a change in the target temperature with time is

$$\dot{X}_E(S^t \mathbf{\Gamma}) = \dot{\beta}(t) H_0(S^t \mathbf{\Gamma}) \tag{8.28}$$

Definition

We defined the *Helmholtz free energy* of the system of interest with temperature T_i in terms of the logarithm of the *canonical partition function* Z_c in Eqs. (5.48) and (5.64). In the present case, we have for the extended system

$$\begin{aligned} A_{Ei} &\equiv -k_{\text{B}} T_i \ln(Z_{cE,i}) \\ &\equiv -k_{\text{B}} T \ln\left(\int d\mathbf{\Gamma} d\alpha\ \exp(-\beta_i H_{E,i}(\mathbf{\Gamma}, \alpha)\right), \quad i = 1, 2 \end{aligned} \tag{8.29}$$

Definition

If we apply the GJE to the canonical case, we see from Eq. (8.9) that, in the canonical case, we obtain an example of the well-known JE:

$$\begin{aligned} \langle \exp[-\Delta X_{c,E,\tau}] \rangle &= \frac{Z_{c,E,2}}{Z_{c,E,1}} = \langle \exp[-\Delta X_{c,\tau}] \rangle \\ &= \frac{Z_{c,2}}{Z_{c,1}} = \exp[-\beta_2 A_2 + \beta_1 A_1] \end{aligned} \tag{8.30}$$

This is a JE for a temperature change in a system with a fixed Hamiltonian and volume.

We also use the fact that $Z_{c,E}(T) = Z_c(T)\sqrt{2\pi/3(N_{\text{th}} - 1)\tau_\alpha^2}$. This allows us to convert to the ratio of partition functions and the free energy differences to those for the system of interest. In the first line, the differences in the extension component of the work also equals zero.

We now use Eq. (8.28) to evaluate the left-hand side of Eq. (8.30). Note that the right-hand side of Eq. (8.28) depends only upon physical variables and not the unphysical thermostat multiplier α or the extended Hamiltonian. Equation (8.30) then becomes

$$\left\langle \exp\left(-\int_0^\tau dt\ \dot{\beta}(t) H_0(S^t\boldsymbol{\Gamma}) \right) \right\rangle_1 = \frac{Z_{c,2}}{Z_{c,1}} = \exp[-\beta_2 A_2 + \beta_1 A_1] \tag{8.31}$$

For quasi-static processes, the exponent of the left-hand side of Eq. (8.31) has no fluctuations, and one can use Eq. (8.31) to show that

$$\begin{aligned} {}_{\text{qs}}\int_0^\tau dt\, \dot{\beta}(t) U(t) &= \beta_2 A_2 - \beta_1 A_1 \\ &=_{\text{qs}} \int_0^\tau dt \frac{d}{dt}[\beta(t)A(t)] \\ &=_{\text{qs}} \int_0^\tau dt\, \dot{T}(t) \frac{\partial}{\partial T}[\beta(t)A(t)] \end{aligned} \tag{8.32}$$

where the subscript "qs" denotes that the integrals are for quasi-static processes only. In deriving Eq. (8.32), we have used the fact that the internal energy U is just the canonical average of the Hamiltonian H_0. For this case, we see that the dimensionless "power" is the rate of change of the dimensionless Helmholtz free energy.

From the second line of Eq. (8.32), we see that for quasi-static processes

$$-\frac{U}{k_B T^2}\dot{T} = -\frac{A}{k_B T^2}\dot{T} + \frac{1}{k_B T}\frac{\partial A}{\partial T}\dot{T} = \frac{d}{dt}(\beta A) \tag{8.33}$$

Multiplying the terms in the first equality by $k_B T^2/\dot{T}$ and rearranging gives, for quasi-static processes

$$A = U + T\frac{\partial A}{\partial T} \tag{8.34}$$

One can, of course, derive this equation directly from the definition of the Helmholtz free energy, Eq. (8.29). Simple differentiation of the Helmholtz free energy with respect to temperature yields Eq. (8.34).

Aside

For temperature changes at finite rates, the classical thermodynamic temperature of the system of interest cannot be defined and the kinetic temperature of the system of interest may not be equal to the temperature of the thermal reservoir. Nonetheless, Eqs. (8.30) and (8.31) can still be used to compute changes in the free energy of the system of interest as specified by Eq. (8.21). In our approach, the temperature is always the temperature of the underlying equilibrium state to which the system will relax if it is so allowed.

If one constructs an algorithm, Eq. (8.21), to accomplish some thermal transformation $(N_1, V_1, T_1) \rightarrow (N_1, V_1, T_2)$, then Eqs. (8.25) and (8.28) give a precise microscopic form for the generalized "work" appearing in the classical thermodynamic path integral for the free energy change. Although the quasi-static path integral expression is unique, the nonequilibrium expression is certainly not. This is because there are infinitely many protocols that accomplish the required change. Nonetheless, each of these expressions gives identical values for the free energy difference.

8.5
Clausius' Inequality, the Thermodynamic Temperature, and Classical Thermodynamics (Evans, Williams, and Searles, 2011)

We now turn our attention away from work, to heat. In the process, we will re-derive the equations of classical thermodynamics we first derived in Section 5.7. However, now we will be able to derive the thermodynamic *inequalities*. As we will soon see, unlike the situation for the thermodynamic *equalities*, the inequalities will, in many cases, be different from what was proposed by Clausius 150 years ago.

As before, we consider a periodic protocol. However, for the heat (and unlike work), we can only deduce useful results if the system responds periodically to the cyclic protocol. We note that, if we periodically cycle a given protocol, not all systems will respond periodically. The necessary and sufficient conditions for the system to respond periodically are not known. Clausius' Inequality for a thermal reservoir applies only if in the long time limit ($t \rightarrow \infty$) the average system response is periodic (Planck, 1945).

Consider a system with the equations of motion given by Eq. (8.21). If we now substitute Eq. (8.25) into Eq. (8.17) and apply it to a *periodic cycle* after any cyclic transients have decayed, we can deduce that

$$\begin{aligned}\lim_{t\rightarrow\infty} \oint_P ds\ \langle \dot{X}(t+s)\rangle &= \lim_{t\rightarrow\infty} \oint_P ds\ \left\langle \beta\,(t+s)\,\dot{Q}_{\text{th}}(t+s) + \frac{d\beta(t+s)H_E(t+s)}{dt}\right\rangle \\ &= \lim_{t\rightarrow\infty} \oint_P \left\langle \frac{dQ_{\text{th}}}{k_B T}\right\rangle \geq 0 \end{aligned} \tag{8.35}$$

where we use the notation $\oint_P ds$ to denote the cyclic integral of a *periodic* function. Because the cycle is periodic, the change in $\langle \beta H_E\rangle$ around the cycle is identically zero. The protocol is, by definition, always periodic, so in one complete cycle $\beta(t+s)$ always returns to its initial value. If and only if the system settles down to a periodic response to the periodic protocol will $\langle H_E(t+s)\rangle$ also return to its initial value at $s = 0$. This can be expected to occur (if it occurs at all) only in the large t limit. In this equation, t is the time that you start the cyclic integral. Not all systems will settle down on average to a periodic response to a periodic protocol.

Thus, for systems that on average settle down to a periodic response to a cyclic protocol, the cyclic integral of the ensemble-averaged change in the heat divided by the target temperature cannot be negative, Eq. (8.35).

In more usual notation, Eq. (8.35) implies that, in the large system limit, that is, $N \to \infty$, where fluctuations are negligible, we obtain the Clausius inequality for the heat transferred to thermal reservoir or a set of thermal reservoirs:

$$\lim_{N\to\infty} \lim_{t\to\infty} \oint_P \frac{dQ_{\text{th}}}{T} \geq 0 \tag{8.36}$$

In these equations, that is, Eqs. (8.35) and (8.36), regardless of the Nosé–Hoover time constant, the time-dependent temperature is the instantaneous value of the *target temperature* of the Nosé–Hoover thermostat. At any instant, the numerical value of the target temperature is, in fact, the equilibrium thermodynamic temperature that the entire system would relax to if at that same moment this target temperature was fixed at its current value and the entire system is allowed to relax to thermodynamic equilibrium. We know that this is so from the relaxation theorem for T-mixing systems (Chapter 5). We will often use the description that the temperature appearing in Eqs. (8.35) and (8.36) is at any instant of time the thermodynamic temperature of the *underlying* equilibrium state.

If the thermostat is composed of a large Hamiltonian region coupled to the system of interest and a remote Nosé–Hoover thermostatted region, we can argue that the precise details of the thermostat cannot possibly be "known" to the system of interest and are therefore unimportant.

If the thermostat is comparable in size to the system of interest and if the cycle is traversed quickly, both the system of interest and the thermostat will be away from equilibrium. At any point in the cycle, there is a profound difference between the nonequilibrium state generated by the Gaussian isokinetic and Nosé–Hoover thermostats. However, for both types of thermostats, Eqs. (8.35) and (8.36) take the same form. For Gaussian thermostats, the change in the kinetic temperature of the thermostat is instantaneous, whereas for Nosé–Hoover thermostats there is a variable phase lag $\sim \tau_\alpha$ in Eq. (8.21). (The value of this feedback time constant is completely arbitrary.) The *only* "temperature" any of these systems have in common is the thermodynamic temperature of the *underlying* equilibrium state. At any point in the cycle, the precise nature of the nonequilibrium state (e.g., the instantaneous average pressure or energy) is highly dependent on the phase lag τ_α or whether the thermostat is Gaussian or Nosé–Hoover-like.

In Planck's discussion of Clausius' inequality (Planck, 1945), at any instant in the cycle, T is the equilibrium thermodynamic temperature of the particular large equilibrium reservoir with which the system of interest is currently in contact.

Clausius' thermodynamic inequality for the reservoirs, Eq. (8.36), is of course exact only in the thermodynamic limit, and in small systems it can occasionally be violated as in Eq. (8.5). The probability ratio that for a finite system the work integral takes on a value A compared to $-A$ can be computed from a time-dependent version of the fluctuation relation.

Equations (8.35) and (8.36) show that, on average, we cannot construct a perpetual motion machine of the second kind. A perpetual motion machine of the second kind would require that $\oint_P \langle dQ_{\text{th}}/T \rangle < 0$ so that ambient heat from the reservoir is converted into useful work in the system of interest. Thus the proof of

Eqs. (8.35) and (8.36) constitutes a direct mechanical proof of Clausius' statement of the second "Law" of thermodynamics.

Because of their importance, we now discuss the case of time reversible quasi-static cycles once again. We already met these in Section 5.7; however, at that time we *only* discussed quasi-static cycles. We also used isokinetic thermostats. Here we use Nosé–Hoover thermostats to obtain basically the same results. We also now have a better understanding of the temperature as it appears in Clausius's equality and his inequality for the thermal reservoirs.

If the cycle is reversible or quasi-static, we can apply Eq. (8.36) to the forward cycle and to the reversed cycle, which must have the same value for the magnitude of the integral but opposite sign. The only possible value for both integrals for reversible cycles is, therefore, zero:

$$\lim_{N\to\infty} {}_{\mathrm{qs}}\oint \frac{đQ_{\mathrm{th}}}{T} = 0 \tag{8.37}$$

The subscript quasi-static denotes a quasi-static cycle. We note that a quasi-static cycle cannot have any transients and is always periodic. This means that we do not need a subscript "p" because it would be redundant. Equation (8.37) is identical to Eq. (5.59), which was derived solely using quasi-static considerations. We note that the strike-through on the differential for the heat, $đQ_{\mathrm{th}}$, denotes the fact that even for quasi-static processes Q is not a state function.

Applying the same arguments as we did for the quasi-static cyclic integral of the generalized work shows that the quasi-static integral from state 1 to state 2

$$\lim_{N\to\infty} {}_{\mathrm{qs}}\int_1^2 \frac{đQ_{\mathrm{th}}}{T} \equiv S_{\mathrm{th},2} - S_{\mathrm{th},1} = -S_{\mathrm{soi},2} + S_{\mathrm{soi},1} \tag{8.38}$$

is independent of the path from state 1 to 2. The subscripts on S indicate that the integral can be evaluated using either the heat transferred to the thermostat or system of interest and $\lim_{N\to\infty\,\mathrm{qs}} \int_1^2 đQ_{\mathrm{soi}}/T \equiv S_{\mathrm{soi},2} - S_{\mathrm{soi},1}$. Using the same arguments as before – see also Eq. (5.60) – the entropy changes for the extended system and for the system of interest itself are identical.

Definition

The quantities $S_{\mathrm{th}}, S_{\mathrm{soi}}$ defined in Eq. (8.38) are termed the *equilibrium* and *calorimetric* entropy of the thermostat and system of interest, respectively. These functions are obviously state functions for quasi-static processes.

If we substitute Eq. (8.25) into Eq. (8.30) using Eq. (8.38) for quasi-static processes in the thermodynamic limit where averages of exponentials equal exponentials of averages, we see that

$$\lim_{N\to\infty} {}_{\mathrm{qs}}\int_i^f \frac{\langle đQ_{\mathrm{th}}\rangle}{T} = -\left(\frac{A_2 - \langle H_{0,2}\rangle}{T_2}\right) + \left(\frac{A_1 - \langle H_{0,1}\rangle}{T_1}\right) = -S_{\mathrm{soi},2} + S_{\mathrm{soi},1} \tag{8.39}$$

In deriving Eq. (8.39), we have used the fact that $\langle \Delta(\beta H_E) \rangle = \langle \Delta(\beta H_0) \rangle$. This particular result was not needed in the isokinetic case – Section 5.7. (Note: we could remove the ensemble averages because in the thermodynamic limit fluctuations vanish.) Now if we compare Eqs. (8.38) and (8.39), we see that in the thermodynamic limit the equilibrium calorimetric entropy S of the system of interest, the Helmholtz free energy A, and the internal energy $\langle H_0 \rangle \equiv U$ must be related by an equation which we have given before, Eq. (5.63):

$$A = \langle H_0 \rangle - TS \equiv U - TS \tag{8.40}$$

We have used the fact that the heat gained by the thermostat is equal and opposite to the heat gained by the system of interest. Comparing Eq. (8.40) with Eq. (8.34) shows that $\partial A/\partial T = -S$. Comparing Eq. (8.40) with Eq. (5.63) shows that (as in Eq. (5.65)) for the quasi-static process:

$$S_{G,c}(T, V) = S(T, V) \tag{8.41}$$

and we see (again) that the Gibbs entropy of an equilibrium canonical ensemble of systems is the same (up to an arbitrary additive constant) as the calorimetric entropy.

We can now also reinterpret Eqs. (8.23)–(8.28) as

$$dU = đQ_{\text{soi}} + dW = TdS + dW \tag{8.42}$$

where all quantities refer to the system of interest. A special case of this Gibbs equation was derived in Eq. (5.62).

Using (8.40), we find that the change in the Helmholtz free energy of the system of interest is given by

$$dA = -SdT + dW \tag{8.43}$$

Finally, we see, from rearranging Eq. (8.40), that

$$S = \frac{U - A}{T} \tag{8.44}$$

which is completely consistent with Eq. (5.63).

Away from equilibrium, we know from Section 2.6 that, if the system is an autonomous Hamiltonian system, the Gibbs entropy of the ensemble is a constant of the motion, and so for these systems one does not have to specify whether the system is at equilibrium or not. We will have more to say of this matter in Section 8.8.

Aside

Equations (8.40)–(8.42) are immediately recognized as the conventional equations of classical thermodynamics for quasi-static processes in the thermodynamic limit. However, unlike classical thermodynamics, our equations came directly from the laws of mechanics and the axiom of causality. We did not need to *assume* Clausius' inequality for the reservoir; we proved it. Indeed, the Helmholtz free energy was defined using the logarithm of the partition function – a statistical mechanical expression rather than a thermodynamic expression. Equation (8.40) is usually taken as the definition of the Helmholtz free energy. In our exposition, we used Eq. (8.29) instead as the definition of the Helmholtz free energy, and Eq. (8.40) is a derived relationship.

Equation (8.38) gives us another very important piece of information. As we saw in Section 5.7, the "integration factor for the heat" in quasi-static processes is the time-dependent thermodynamic temperature. The fact that the equilibrium temperature is the integrating factor for the heat ultimately comes from the form of the canonical equilibrium distribution function – see Section 5.7 for more details. The equilibrium relaxation theorem (Chapter 5) says that this distribution is *unique* for T-mixing systems. The consequence of this is that the integrating factor for the heat, that is, T in Eq. (8.38), is also *unique*.

We have now seen that our mechanical derivations of the fundamental *equality* of classical thermodynamics is independent of whether we use isokinetic dynamics as in Section 5.7 or Nosé–Hoover dynamics as in this section. In this section we have also been able to derive the Clausius inequality for the reservoir. In our derivation, we have been able to give meaning to the temperature appearing in this inequality as it applies to nonequilibrium cycles. This avoids the logical problems with the Clausius inequality that were first raised by Bertrand (1887), Orr (1904), and Buckingham (1905). Even if the system is far from equilibrium during the cycle (where Clausius' temperature becomes ill-defined), the temperature that appears in the Clausius inequality refers to a well-defined equilibrium temperature.

In our "Aside" on the "The Thermodynamic Connection" in Section 5.6, we proved that if we *assume* the traditional "laws" of thermodynamics, the thermodynamic Helmholtz free energy and the microscopic partition function are related by Eq. (5.48) (which is identical to Eq. (8.29)). We also proved for equilibrium systems the equivalence of the target kinetic temperature employed in a Nosé–Hoover thermostat and the thermodynamic temperature for canonical systems. These proofs took the classical thermodynamic relations between the Helmholtz free energy, the internal energy, the entropy, volume, and temperature as *given*. Section 5.6 was labeled as an "Aside" because it is not necessary for the logical exposition of this book.

We have now arrived at a completely new logical position. We have *proved* the zeroth (Section 7.2) and second "Laws" of thermodynamics (Section 8.5), the latter in the form of the Clausius inequality for the reservoir, Eqs. (8.35) and (8.36). We take the first "Law" of thermodynamics as given by the laws of mechanics. This means that logically we have now proved the macroscopic equations that are the basis of thermodynamics, Eqs. (8.40)–(8.42). We made no assumptions except the laws of mechanics, the assumption of T-mixing, ergodic consistency, and the axiom of causality. Clausius, of course, proved his theorem *assuming* the second "Law" of thermodynamics. He *assumed* that the construction of a perpetual motion machine of the second kind was impossible. Our proof requires no such assumption. This is quite a different logical point of view from that used by Clausius (1854).

We take the mechanical quantities such as energy and pressure as microscopic averages of the mechanical energy and pressure. We *define* the Helmholtz free energy in terms of the logarithm of the canonical partition function. We proved Clausius' inequality for the cyclic integral of the heat divided by the temperature of the underlying equilibrium state. The temperature is just a parameter we met in the unique equilibrium phase space distribution that nonequilibrium canonical systems relax toward (Section 5.4). Clausius' equality for the heat allowed us to define entropy as a state function for equilibrium or quasi-static systems. This then allowed us to generate Eq. (8.41), which is the Gibbs equation that expresses the thermodynamic summary of the first and second "Laws" of thermodynamics as they apply to quasi-static processes. Knowing the relationship between the internal and Helmholtz free energies allowed us to derive Gibbs' *microscopic* expression for the equilibrium thermodynamic or calorimetric entropy. At no stage in our exposition did we need to assume the "Laws" of thermodynamics. They were derived.

8.6 Purely Dissipative Generalized Work

Definition

We define a *dimensionless, purely dissipative generalized "work"* $\Delta Y_\tau(\mathbf{\Gamma})$ for a trajectory of duration τ, originating from the phase point $\mathbf{\Gamma}$, under this dynamics as (Reid, Sevick, and Evans, 2005)

$$\exp[\Delta Y_\tau(\mathbf{\Gamma})] = \frac{p_{\text{eq},1}(d\mathbf{\Gamma})}{p_{\text{eq},2}(d(S^\tau\mathbf{\Gamma}))} \equiv \frac{f_{\text{eq},1}(\mathbf{\Gamma})d\mathbf{\Gamma}}{f_{\text{eq},2}(S^\tau\mathbf{\Gamma})d(S^\tau\mathbf{\Gamma})}, \quad \forall\mathbf{\Gamma} \in D \tag{8.45}$$

The derivation of results for this quantity is very similar to that for the generalized dimensionless work (Sections 8.1–8.3), so we will quickly give a summary of the main results without rehearsing the proofs.

As before, the time τ is the time at which the parametric change in λ is complete. This means that at time τ this system is *not* necessarily at equilibrium.

Following the same procedure that led to Eq. (8.10), we see that

$$\langle\exp[-\Delta Y_\tau(\mathbf{\Gamma})]\rangle_{\text{eq},1} = \int d\mathbf{\Gamma} f_{\text{eq},1}(\mathbf{\Gamma};0)\frac{f_{\text{eq},2}(S^\tau\mathbf{\Gamma};0)d(S^\tau\mathbf{\Gamma})}{f_{\text{eq},1}(\mathbf{\Gamma};0)d\mathbf{\Gamma}} = 1 \tag{8.46}$$

where the brackets $\langle\cdots\rangle_{\text{eq},1}$ denote an equilibrium ensemble average over the initial distribution. This equation is the analog of the NPI.

The validity of Eq. (8.46) requires that there is an integrable region in the phase space of the final equilibrium distribution for which $f_{\text{eq},2}(S^\tau\mathbf{\Gamma};0) \neq 0$, that is, $\int d(S^\tau\mathbf{\Gamma})\ f_{\text{eq},2}(S^\tau\mathbf{\Gamma};0) \neq 0$.

Definition

The requirement that for any $f_{\text{eq},1}(\mathbf{\Gamma};0) \neq 0$ the existence of an integrable region in the phase space of the final equilibrium distribution for which $f_{\text{eq},2}(S^\tau\mathbf{\Gamma};0) \neq 0$, that is, $\int d(S^\tau\mathbf{\Gamma})\ f_{\text{eq},2}(S^\tau\mathbf{\Gamma};0) \neq 0$, is the *ergodic consistency condition for the generalized work.*

The fluctuation relation for the purely dissipative generalized work considers the probability $p_{\text{eq},1}(\Delta Y_\tau = B)$ of observing values of $\Delta Y_\tau = B \pm dB$ for forward trajectories starting from the initial equilibrium distribution $f_{\text{eq},1}(\mathbf{\Gamma},0)$, and the probability $p_{\text{eq},2}(\Delta Y_\tau = -B)$ of observing $\Delta Y_\tau = -B \pm dB$ for reverse trajectories but starting from the equilibrium given by $f_{\text{eq},2}(\mathbf{\Gamma},0)$. Proof of this GCFR closely resembles those of ESFT and the GCFT Eq. (8.6):

$$\begin{aligned}
\frac{p_{\text{eq},1}(\Delta Y_{\tau,f} = B \pm dB)}{p_{\text{eq},2}(\Delta Y_{\tau,r} = -B \mp dB)} &= \frac{\displaystyle\int_{\Delta Y_{\tau,f}(\mathbf{\Gamma})=B\pm dB} d\mathbf{\Gamma} f_{\text{eq},1}(\mathbf{\Gamma})}{\displaystyle\int_{\Delta Y_{\tau,r}(\mathbf{\Gamma})=-B\mp dB} d\mathbf{\Gamma} f_{\text{eq},2}(\mathbf{\Gamma})} \\
&= \frac{\displaystyle\int_{\Delta Y_{\tau,f}(\mathbf{\Gamma})=B\pm dB} d\mathbf{\Gamma} f_{\text{eq},1}(\mathbf{\Gamma})}{\displaystyle\int_{\Delta Y_{\tau,r}(M^T S^\tau\mathbf{\Gamma})=-B\mp dB} dM^T S^\tau\mathbf{\Gamma} f_{\text{eq},2}(M^T S^\tau\mathbf{\Gamma})} \\
&= \frac{\displaystyle\int_{\Delta Y_{\tau,f}(\mathbf{\Gamma})=B\pm dB} d\mathbf{\Gamma} f_{\text{eq},1}(\mathbf{\Gamma})}{\displaystyle\int_{\Delta Y_{\tau,f}(\mathbf{\Gamma})=B\pm dB} d\mathbf{\Gamma}\ \exp[-\Delta Y_{\tau,f}(\mathbf{\Gamma})] f_{\text{eq},1}(\mathbf{\Gamma})} \\
&= \exp[B]
\end{aligned} \tag{8.47}$$

Again, we can see the change in the purely dissipative work *after* the parametric changes are complete:

$$\exp[\Delta Y_{\tau+s}(\mathbf{\Gamma}) - \Delta Y_\tau(\mathbf{\Gamma})] = \frac{f_{\text{eq},2}(S^\tau\mathbf{\Gamma})d(S^\tau\mathbf{\Gamma})}{f_{\text{eq},2}(S^{\tau+s}\mathbf{\Gamma})d(S^{\tau+s}\mathbf{\Gamma})} \tag{8.48}$$

which is recognized as the dissipation function for the second equilibrium. Therefore, we have

$$\begin{aligned}
[\Delta Y_{\tau+s}(\mathbf{\Gamma}) - \Delta Y_\tau(\mathbf{\Gamma})] &\equiv \Delta Y_s(S^\tau\mathbf{\Gamma}) \\
&= \Omega_{\text{eq},2,s}(S^\tau\mathbf{\Gamma}) = 0, \quad \forall\mathbf{\Gamma} \in D,\ \forall s > 0
\end{aligned} \tag{8.49}$$

This leads to the shortened form for the fluctuation relation for the dimensionless purely dissipative generalized work:

$$\frac{p_{\text{eq},1}(\Delta Y_\tau = B)}{p_{\text{eq},2}(\Delta Y_\tau = -B)} = \exp[B] \tag{8.50}$$

The derivation of the JE for the dimensionless, purely dissipative generalized work is trivial:

$$\begin{aligned}\langle \exp[-\Delta Y_\tau] \rangle_{\mathrm{eq},1} &= \int_{D_1} d\mathbf{\Gamma}\ f_{\mathrm{eq},1}(\mathbf{\Gamma}) \frac{f_{\mathrm{eq},2}(S^\tau \mathbf{\Gamma}) d(S^\tau \mathbf{\Gamma})}{f_{\mathrm{eq},1}(\mathbf{\Gamma}) d\mathbf{\Gamma}} \\ &= \int_{D_2} d(S^\tau \mathbf{\Gamma}) f_{\mathrm{eq},2}(S^\tau \mathbf{\Gamma}) \\ &= 1 \end{aligned} \tag{8.51}$$

Now we are in a position to give a simple but informative derivation of the GJE. Comparing Eq. (8.1) with Eq. (8.45), we see that the generalized work is related to the purely dissipative work by the relation

$$\exp[\Delta X_\tau(\mathbf{\Gamma})] = \exp[\Delta Y_\tau(\mathbf{\Gamma})] \frac{Z_1}{Z_2} \tag{8.52}$$

This means that

$$\begin{aligned}\langle \exp[-\Delta X_\tau] \rangle_{\mathrm{eq},1} &= \langle \exp[-\Delta Y_\tau] \rangle_{\mathrm{eq},1} \frac{Z_2}{Z_1} \\ &= \frac{Z_2}{Z_1} \end{aligned} \tag{8.53}$$

which is identical to the GJE, Eq. (8.9). For isothermal processes, this becomes rather simple: $\Delta X_\tau \to \beta \Delta W_\tau$, $\Delta Y_\tau \to \beta \Delta W_{\mathrm{irr},\tau}$, and $\beta \Delta W_\tau = \beta \Delta W_{\mathrm{irr},\tau} + \beta \Delta A$.

The generalized, purely irreversible work $\Delta Y_{irr,\tau}$ has properties very similar to those of the dissipation function. However, there are important differences. States 1 and 2 must be in equilibrium, and Eq. (8.45) refers to forward and reverse trajectories whereas the dissipation function (as in Eq. (8.49)) only refers to forward processes.

Nevertheless, the dissipative work shares many analogous properties. From the JE it is easy to compute a bound on the purely dissipative, dimensionless, generalized work for a thermodynamic process (Jarzynski, 1997a):

$$\begin{aligned} 1 &= \langle \exp[-\Delta Y_\tau] \rangle_{\mathrm{eq},1} \\ &= \exp[-\langle \Delta Y_\tau \rangle_{\mathrm{eq},1}] \langle \exp[-\Delta Y_\tau + \langle \Delta Y_\tau \rangle_{\mathrm{eq},1}] \rangle_{\mathrm{eq},1} \\ &\geq \exp[-\langle \Delta Y_\tau \rangle_{\mathrm{eq},1}] \langle 1 - \Delta Y_\tau + \langle \Delta Y_\tau \rangle_{\mathrm{eq},1} \rangle_{\mathrm{eq},1} = \exp[-\langle \Delta Y_\tau \rangle_{\mathrm{eq},1}] \end{aligned} \tag{8.54}$$

In deriving this result we have used the fact that $e^x \geq 1 + x, \quad \forall x$. The above equation implies that the ensemble average of the purely irreversible dimensionless work is positive except for quasi-static processes:

$$\langle \Delta Y_\tau \rangle_{\mathrm{eq},1} \geq 0 \tag{8.55}$$

This is formally analogous to the second law inequality. If a process is reversible, the change in the work for the forward path must be equal and opposite to that for the reverse path and the only way this can occur is if the change in the work is zero. This shows that the work is *purely* dissipative, as claimed when it was defined. For quasi-static averages

$$\langle \Delta Y \rangle_{\mathrm{qs}} = 0 \tag{8.56}$$

In spite of these similarities between the purely irreversible work and the dissipation function, their respective definitions show that they are quite different. This is also evidenced by the fact that the proof of the second law inequality for the purely irreversible work has to be different from that for the dissipation function.

8.7
Application of the Crooks Fluctuation Theorem (CFT), and the Jarzynski Equality (JE)

We now give an example of how to apply the GCFT and GJE to an actual statistical mechanical ensemble and system of dynamics. We show that these very general results lead to the canonical forms of the CFT and the JE for the transformation between initial and final equilibrium states with the same values for temperature, volume, and number of particles (T, V, N). In Section 8.4, we already met the JE for a canonical system subject to a changing temperature. Here we give the JE and Crooks fluctuation relations for Nosé–Hoover isothermal systems in which the Hamiltonian is subject to a parametric transformation. This is the usual transformation that is discussed by these relations and was the original relation proposed by Jarzynski (1997a).

We assume all systems are T-mixing over the respective phase space domains. The relevant equilibrium distribution function is the canonical distribution function – see Section 5.3

$$f(\mathbf{\Gamma};0) = \frac{\exp[-\beta H_0(\mathbf{\Gamma})]}{Z_c}, \quad \forall \mathbf{\Gamma} \in D \tag{8.57}$$

In Eq. (8.28), we defined a quantity called the *Helmholtz free energy, A*, which is related to logarithm of the canonical partition function Z_c – see Sections 5.3, 5.4, and 5.6,

$$\begin{aligned} A(\lambda) &\equiv -k_B T \ln Z_c(\lambda) \\ &= -k_B T \ln \left[\int d\mathbf{\Gamma} \exp\left(-\beta H_0 \left(\mathbf{\Gamma}, \lambda\right)\right) \right] \end{aligned} \tag{8.58}$$

In order to transform from the initial equilibrium state with $\lambda = \lambda_1 = \lambda(0)$ to the final equilibrium state with $\lambda = \lambda_2 = \lambda(\tau)$, the functional form of the system's Hamiltonian may vary parametrically over the period $0 < t < \tau$, for example, $H_0(\mathbf{\Gamma}, \lambda(t)) = \sum_{i=1}^{N} p_i^2/(2m) + \Phi(\mathbf{q}, \lambda(t))$, where $\Phi(\mathbf{q}, \lambda(t))$ is the interparticle potential. For $t > \tau$, the Hamiltonian's parametric dependence is fixed at $H_0(\mathbf{\Gamma}, \lambda(\tau))$. Over the time $0 < t < \tau$, the ensemble is driven arbitrarily far away from equilibrium, and if the transformation is halted at $t = \tau$, because the system is T-mixing, the system will eventually relax to the unique, new equilibrium state.

For a system in which the phase space is extended because of the introduction of additional dynamical variables such as the volume or those associated with the thermostat (such as in the case of Nosé–Hoover dynamics (Hoover, 1985), as detailed below), the work becomes $\Delta X_\tau = \beta[H_E(S^\tau\mathbf{\Gamma}, \lambda(\tau)) - H_E(\mathbf{\Gamma}, \lambda(0))] + \ln\left[\frac{d\mathbf{\Gamma}}{d(S^\tau\mathbf{\Gamma})}\right]$ where H_E is the Hamiltonian of the extended system (Williams, Searles, and Evans, 2008). We note, as before (Section 8.4), that from the equations of motion for the Nosé–Hoover thermostat Eq. (8.21), the differences in the extended Hamiltonian divided by $k_B T$ are the same as for the internal energy because the extended Hamiltonian divided by $k_B T$ differs from the internal energy divided by $k_B T$ by terms that are independent of the temperature and the Hamiltonian of the system of interest. The difference is, in fact, $3(N_{\text{th}} - 1)\alpha^2\tau^2/2$. Alternatively, we could more simply employ isokinetic dynamics, thereby obviating the need for an extension to the Hamiltonian. The results are formally identical.

Using Eq. (8.1), the generalized "work" becomes

$$\begin{aligned}
\Delta X_\tau &= \beta[H_0(S^\tau\mathbf{\Gamma}, \lambda(\tau)) - H_0(\mathbf{\Gamma}, \lambda(0))] + \ln\left[\frac{d\mathbf{\Gamma}}{d(S^\tau\mathbf{\Gamma})}\right] \\
&= \beta[H_0(S^\tau\mathbf{\Gamma}, \lambda(\tau)) - H_0(\mathbf{\Gamma}, \lambda(0))] - \int_0^\tau ds\ \Lambda(S^s\mathbf{\Gamma}) \\
&= \beta[H_0(S^\tau\mathbf{\Gamma}, \lambda(\tau)) - H_0(\mathbf{\Gamma}, \lambda(0)) + \Delta Q_\tau] \\
&= \beta\Delta W_\tau
\end{aligned} \tag{8.59}$$

The final equality is obtained from the first "Law" of thermodynamics, and the equations of motion must satisfy $AI\mathbf{\Gamma}$.

Definition

So the generalized dimensionless "work" is, in the isothermal canonical case, identifiable as β times the *work* performed over a period of time τ. The latter is the change in energy *minus* the change in energy due solely to the exchange of heat: $\Delta W = \Delta U + \Delta Q_{\text{th}} = \Delta U_{\text{soi}} - \Delta Q_{\text{soi}}$. (Remember, $\dot{Q}_{\text{th}}$ is defined, Eq. (8.23), as the heat transferred to the thermostat, which is the negative of the heat transferred to the system.)

We note that, if at the end of the protocol $t = \tau$ the system is not in equilibrium, it does not matter. Any subsequent relaxation processes will have *no* effect on ΔW. The change in the energy is *exactly* due to the change in the heat, leaving the work unchanged, exactly as proved for the generalized work in Eq. (8.6).

Using Eqs. (8.6) and (8.59), the CFT is given as

$$\frac{p_1(\Delta W_\tau = B)}{p_2(\Delta W_\tau = -B)} = \exp[\beta B]\frac{Z_{c,2}}{Z_{c,1}} = \exp[-\beta(\Delta A - B)] \tag{8.60}$$

where $\Delta A = A_2 - A_1 = A(\lambda(\tau)) - A(\lambda(0))$, and using Eq. (8.10), the JE is

$$\langle \exp(-\beta\Delta W_\tau)\rangle = \frac{Z_{c,2}}{Z_{c,1}} = \exp(-\beta\Delta A) \tag{8.61}$$

The same results are obtained for the canonical distribution when the dynamics are thermostatted by a Gaussian thermostat (Evans, 2003) or a Nosé–Hoover thermostat (Williams, Searles, and Evans, 2008). For other ensembles and transformations, Eq. (8.1) does not necessarily refer to a work (e.g., see Reid, Sevick, and Evans (2005), Adib (2005), and Chelli *et al.* (2007)).

The first experimental tests of the JE and CFR were by Liphardt *et al.* (2002), who used optical tweezers to extend a DNA–RNA hybrid chain, measuring the work required as the extension proceeded.

If we now evaluate the cyclic work integral Eq. (8.17) for the case of constant temperature systems, we have

$$\begin{aligned}\oint \langle dW \rangle &= \oint \langle dU_{\text{soi}} \rangle - \oint \langle dQ_{\text{soi}} \rangle \\ &= \oint \langle dU_{\text{soi}} \rangle + \oint \langle dQ_{\text{th}} \rangle \geq 0 \end{aligned} \tag{8.62}$$

where, as usual, the cycle is defined as a cycle in the parameter $\lambda(t)$. The validity of Eq. (8.62) is independent of whether the system responds periodically to the cycle. It only requires that the parameter $\lambda(s)$ returns to its initial value. The system at the end of this cycle may have an internal energy that, on average, is different from its average initial value. This is because, if the cycle begins from an equilibrium state but is traversed quickly, then at the end of the cycle the system may be in a nonequilibrium state with an internal energy that is different from the initial equilibrium value.

When the parameter completes a cycle, $\lambda(\tau) = \lambda(0)$, the subsequent change in the internal energy dU_{sys} is identical to the heat absorbed by the system dQ_{soi} from the thermostat since there is no further change in the work. So even though at the end of a parametric cycle the system is not yet in equilibrium and subsequent thermal relaxation will still take place, the CI, Eq. (8.27), is still valid.

If we run the cycle in reverse (i.e., we reverse the direction of the protocol) and if the process is reversible, then the cyclic integral will take on the opposite sign (by definition) but still obey the inequality, Eq. (8.26). The only way this can be true is if for reversible systems the cyclic integral of the work is zero.

Definition

We call Eq. (8.62) the *nonequilibrium cyclic work inequality* for the system of interest.

If and only if the cycle is *periodic* (i.e., the system has periodically cycled many times through a periodic protocol and *if* the initial transients have decayed to zero), then $\oint_P \langle dU_{\text{soi}} \rangle = 0$, and then Eq. (8.62) gives

$$\oint_P \langle dQ_{\text{th}} \rangle \geq 0 \tag{8.63}$$

Further, if we now take the thermodynamic limit, we see that for large systems the heat absorbed by a large heat reservoir over one periodic cycle cannot be negative. If the process is conducted reversibly, then both the forward periodic cycle and the reverse periodic cycle must both be nonnegative and therefore both must integrate to zero.

In summary, if we surround our system of interest with a *large* equilibrium thermal reservoir at temperature T, and if we do a cycle of work defined by some protocol $\lambda(t)$, we see that over the cycle, on average, the thermostat absorbs heat rather than losing it.

In small systems where the cycles are of limited duration, there will be instances when $\oint dW < 0$! Our result says that *on average* the cyclic inequality for the work holds.

8.8 Entropy Revisited

"Entropy is not conserved; it is increasing all the time." (Penrose, 1990, p. 412)

We now look again at Eq. (8.35). If our system is subject to a periodic thermal protocol and if the system settles into a periodic cycle, the ensemble-averaged heat absorbed by the thermostat (dQ_{th}) is nonnegative, $\lim_{t\to\infty} \oint_P \langle dQ_{\mathrm{th}}/T\rangle \geq 0$. If the sign were reversed, we would have been able to construct a perpetual motion machine of the second kind. So we have given a proof of the second "Law" of thermodynamics since Clausius' statement of that "Law" refers to the impossibility of constructing such a machine.

There is a complementary inequality for the system of interest, namely

$$\lim_{t\to\infty} \oint_P \langle dQ_{\mathrm{soi}}/T\rangle \leq 0 \tag{8.64}$$

If we combine the system of interest and the thermostat, we see that for the combined system, the "Universe," we find that for both reversible and irreversible processes

$$\begin{aligned} dQ_{\mathrm{soi}}(t)/T(t) + dQ_{\mathrm{th}}(t)/T(t) &= (dQ_{\mathrm{soi}}(t) - dQ_{\mathrm{soi}}(t))/T(t) \\ &= 0, \quad \forall t \end{aligned} \tag{8.65}$$

The second line follows because, first, $dQ_{\mathrm{soi}}(t) = -dQ_{\mathrm{th}}(t)$ and, second, at any point in the cycle the temperature of the underlying equilibrium states for both the thermostat and the system of interest are, by the zeroth law, equal because both systems, by construction, are in thermal contact. Defining the change in the entropy of the system and thermostat by Eq. (8.38) and the change in entropy of the Universe as a sum of these, Eq. (8.65), directly contradicts Clausius's assertion (Clausius, 1854) that the entropy of the Universe tends to a maximum. It is dissipation that tends to a maximum, not entropy. This result is a direct consequence of the definition of temperature in Eq. (8.38). If, instead, the average kinetic temperature of the system/thermostat was used in definition of the change in entropy and the temperature of the thermostat was increasing, the kinetic temperature of the system would be less than that of the thermostat. Then the total change in entropy would be positive. However, for a process that is not quasi-static, the temperature in Clausius's original work was not defined. There are infinitely many definitions of the temperature of a nonequilibrium system; it is the underlying equilibrium temperature that naturally occurs in the second law inequality from which Eq. (8.36) is obtained. Equation (8.65) also avoids the criticisms of Clausius' inequality, as it applies to irreversible processes (Bertrand, 1887; Orr, 1904; Buckingham, 1905). The temperature appearing in Eqs. (8.36) and (8.8.2) is well defined even for irreversible processes.

In spite of our knowledge of the time dependence of the entropy in nonequilibrium systems, away from equilibrium entropy is not a useful quantity. Away from equilibrium it is dissipation that is useful. At equilibrium, their respective roles are reversed because, at equilibrium, dissipation is identically zero while entropy becomes exceedingly useful.

8.9 For Thermostatted Field-Free Systems, the Nonequilibrium Helmholtz Free Energy is a Constant of the Motion

Consider a thermostatted system in contact with an isokinetic heat bath. The bath could be much larger than the system of interest, in which case the heat bath could be approximated as being in thermodynamic equilibrium while the system of interest, which is in thermal contact with the bath, relaxes toward equilibrium. The heat bath could also be of similar size to the system of interest, and therefore it may also be out of equilibrium.

There are no external dissipative fields applied to the system, but the initial distribution for the system of interest is not an equilibrium distribution. The nonequilibrium system of interest is relaxing toward equilibrium. In Eqs. (2.56) and (5.66), we saw there that the rate of change of the Gibbs entropy and the irreversible calorimetric entropy is given by the equation

$$\dot{S}_G(t) = \dot{S}_{\text{ir}}(t) = k_{\text{B}} \int d\mathbf{\Gamma} f(\mathbf{\Gamma}; t)) \frac{\partial}{\partial \mathbf{\Gamma}} \cdot \dot{\mathbf{\Gamma}} = -k_{\text{B}}(3N - 4)\langle \alpha(t) \rangle \tag{8.66}$$

where N_{th} is the number of thermostatted particles, and α is the usual isokinetic thermostat multiplier. We also know that the rate of change of the total internal energy of the system of interest is

$$\begin{aligned} \langle \dot{H}_0(t) \rangle &= -2K_{\text{th}} \langle \alpha(t) \rangle \\ &\equiv -3N_{\text{th}} k_{\text{B}} T \langle \alpha(t) \rangle \end{aligned} \tag{8.67}$$

where T is the kinetic temperature of the reservoir, which is, of course, also equal to the thermodynamic temperature of the underlying equilibrium system.

Since the system is not initially at equilibrium, we cannot use the equilibrium statistical mechanical definition of the Helmholtz free energy, which for a canonical equilibrium state is given by Eq. (5.64): $A \equiv -k_{\text{B}} T \ln \left[\int d\mathbf{\Gamma}\, e^{-\beta H_0} \right]$. Instead, we can use Eq. (8.40) to define the Helmholtz free energy of the nonequilibrium system:

$$A_{\text{ne}} \equiv U - TS_{\text{ir}} \tag{8.68}$$

From Eqs. (8.66)–8.68, we can then deduce that the nonequilibrium Helmholtz free energy for the system of interest is a constant of the motion:

$$\dot{A}_{\rm ne} = \langle \dot{H}_0 \rangle - T\dot{S}_{\rm ir} = 0 \tag{8.69}$$

If the initial perturbed system was a canonical equilibrium distribution with a temperature that was different from that of the thermostat, and the system was allowed to relax to the new equilibrium state, then Eq. (5.64) could be used to determine the difference in the equilibrium Helmholtz free energies of the two equilibrium states. Unlike the difference in the nonequilibrium Helmholtz free energies of the two states, the difference in the equilibrium Helmholtz free energies will be nonzero.

These systems, with an initial distribution that is not the equilibrium distribution of the dynamics, are precisely the type of systems treated by the relaxation theorem for canonical systems in contact with a heat reservoir. In that case, however, the time integral of the ensemble average of the dissipation is positive for all times, and the instantaneous dissipation eventually decays toward zero everywhere in phase space (i.e., in the infinite time limit, the system must have relaxed to equilibrium). Dissipation consistently describes the nonequilibrium relaxation process. In contrast, Eq. (8.69) is an example which shows that away from equilibrium free energies like the Helmholtz free energy are, like the entropy, of little use.

References

Adib, A.B. (2005) Entropy and density of states from isoenergetic nonequilibrium processes. *Phys. Rev. E*, **71**, 056128.

Adjanor, G., Athènes, M., and Calvo, F. (2006) Free energy landscape from path-sampling: application to the structural transition in Lj_{38}. *Eur. Phys. J. B*, **53**, 47–60.

Bertrand, J.L.F. (1887) *Thermodynamique*, Gauthier-Villars, Paris.

Buckingham, E. (1905) On certain difficulties which are encountered in the study of thermodynamics. *Philos. Mag.*, **9**, 208.

Chelli, R., Marsili, S., Barducci, A., and Procacci, P. (2007) Generalization of the Jarzynski and Crooks nonequilibrium work theorems in molecular dynamics simulations. *Phys. Rev. E*, **75**, 050101.

Clausius, R. (1854) Uber Eine Veranderte Form Des Zweiten Hauptsatzes Der Mechanischen Warmtheorie. *Ann. Phys. Chem.*, **93**, 481.

Clausius, R. (1856) On a modified form of the second fundamental theorem in the mechanical theory of heat. *Philos. Mag.*, **4**, 12.

Crooks, G.E. (1999) Entropy production fluctuation theorem and the nonequilibrium work relation for free energy differences. *Phys. Rev. E: Stat. Phys. Plasmas Fluids Relat. Interdiscip. Top.*, **60**, 2721–2726.

Evans, D.J. (2003) A non-equilibrium free energy theorem for deterministic systems. *Mol. Phys.*, **101**, 1551–1554.

Evans, D.J., Williams, S.R., and Searles, D.J. (2011) A proof of Clausius' theorem for time reversible deterministic microscopic dynamics. *J. Chem. Phys.*, **134**, 204113-1.

Hoover, W.G. (1985) Canonical dynamics: equilibrium phase-space distributions. *Phys. Rev. A*, **31**, 1695–1697.

Jarzynski, C. (1997a) Equilibrium free-energy differences from nonequilibrium measurements: a master-equation approach. *Phys. Rev. E*, **56**, 5018–5035.

Jarzynski, C. (1997b) Nonequilibrium equality for free energy differences. *Phys. Rev. Lett.*, **78**, 2690–2693.

Kofke, D.A. (2006) On the sampling requirements for exponential-work free-energy calculations. *Mol. Phys.*, **104**, 3701–3708.

Lechner, W. and Dellago, C. (2007) On the efficiency of path sampling methods for the calculation of free energies from non-equilibrium simulations. *J. Stat. Mech.-Theory Exp.*, **2007**, P04001.

Lechner, W., Oberhofer, H., Dellago, C., and Geissler, P.L. (2006) Equilibrium free energies from fast-switching trajectories with large time steps. *J. Chem. Phys.*, **124**, 044113.

Liphardt, J., Dumont, S., Smith, S.B., Tinoco, I., and Bustamante, C. (2002) Equilibrium information from nonequilibrium measurements in an experimental test. *Science*, **296**, 1832–1835.

Lua, R.C. and Grosberg, A.Y. (2005) Practical applicability of the Jarzynski relation in statistical mechanics: a pedagogical example. *J. Phys. Chem. B*, **109**, 6805–6811.

Macfadyen, J. and Andricioaei, I. (2005) A skewed-momenta method to efficiently generate conformational-transition trajectories. *J. Chem. Phys.*, **123**, 074107.

Maxwell, J.C. (1878) Tait's "thermodynamics" li. *Nature*, **17**, 278–280.

Orr, W.M. (1904) On Clausius's theorem for irreversible cycles, and on the increase of entropy. *Philos. Mag. Ser. 6*, **8**, 509.

Penrose, R. (1990) *The Emperor's New Mind*, Vantage, London.

Planck, M. (1945) *Treatise on Thermodynamics*, Dover, Mineola, NY.

Reid, J.C., Sevick, E.M., and Evans, D.J. (2005) A unified description of two theorems in non-equilibrium statistical mechanics: the fluctuation theorem and the work relation. *Europhys. Lett.*, **72**, 726–732.

Schmiedl, T. and Seifert, U. (2007) Optimal finite-time processes in stochastic thermodynamics. *Phys. Rev. Lett.*, **98**, 108301.

Shirts, M.R. and Pande, V.S. (2005) Comparison of efficiency and bias of free energies computed by exponential averaging, the Bennett acceptance ratio, and thermodynamic integration. *J. Chem. Phys.*, **122**, 144107.

Vaikuntanathan, S. and Jarzynski, C. (2008) Escorted free energy simulations: improving convergence by reducing dissipation. *Phys. Rev. Lett.*, **100**, 190601.

Williams, S.R., Searles, D.J., and Evans, D.J. (2008) Nonequilibrium free-energy relations for thermal changes. *Phys. Rev. Lett.*, **100**, 250601.

Wu, D. and Kofke, D.A. (2005a) Phase-space overlap measures. I. Fail-safe bias detection in free energies calculated by molecular simulation. *J. Chem. Phys.*, **123**, 054103.

Wu, D. and Kofke, D.A. (2005b) Phase-space overlap measures. II. Design and implementation of staging methods for free-energy calculations. *J. Chem. Phys.*, **123**, 084109.

Wu, D. and Kofke, D.A. (2005c) Rosenbluth-sampled nonequilibrium work method for calculation of free energies in molecular simulation. *J. Chem. Phys.*, **122**, 204104.

Ytreberg, F.M., Swendsen, R.H., and Zuckerman, D.M. (2006) Comparison of free energy methods for molecular systems. *J. Chem. Phys.*, **125**, 184114.

9
Causality

> In quantum mechanics … The fundamental equation is itself symmetrical under time reversal … However, despite this symmetry, quantum mechanics does in fact involve an important non-equivalence of the two directions of time. This appears in connection with the interaction of a quantum object with a system which with sufficient accuracy obeys the laws of classical mechanics … If two interactions A and B with a given quantum object occur in succession, then the statement that the probability of any particular result of process B is determined by the result of process A can be valid only if process A occurred earlier than process B.
>
> *(Landau and Lifshitz, 1969, p. 31)*

9.1
Introduction

If *all* the laws of mechanics and quantum mechanics were time-reversal-symmetric, then clearly you cannot prove a time-asymmetric result like the fluctuation theorem. In the first proof given by Evans and Searles (1994), this time symmetry was indeed broken, but it was broken in such a natural way that most people who have analyzed this proof fail to see where the time reversal symmetry is broken. Time reversal symmetry was broken when it was assumed that natural processes are *causal* (Searles and Evans, 1996).

We quote Landau and Lifshitz above (p. 31). This is a statement of the axiom of *causality* at least as it applies to quantum mechanics. It is used frequently in quantum mechanics, but (not indicated by Landau and Lifshitz) it is also required in classical mechanics and electrodynamics. The equations of motion in classical (and quantum) mechanics are indifferent to the direction of time – Hamilton's action principle shows this with great clarity. However, mechanics on its own does not give us enough information to predict experimental results.

In the proofs of the ESFT (Evans – Searles transient fluctuation theorem) and the GCFT (generalized Crooks fluctuation theorem), the probabilities of observing particular values of time integrals of the dissipation function or of the generalized work are computed from the probabilities of observing the *initial* states

Fundamentals of Classical Statistical Thermodynamics: Dissipation, Relaxation and Fluctuation Theorems, First Edition. Denis J. Evans, Debra J. Searles, and Stephen R. Williams.

from which those sets of trajectories *began:* $f(\mathbf{\Gamma};0)d\mathbf{\Gamma}$. We never computed the probabilities of observing conjugate sets of trajectories from the known endpoint distribution; indeed, had we done so, we would have proved the "anti-fluctuation theorem" and an "anti-second law" (Evans and Searles, 1996).

What is so strange about the anti-causal Universe is that in this case the phase space distribution and the last time (t) determine the initial distribution of phases at $t = 0$ for all values of t! For nonequilibrium antisteady states in an anti-causal Universe, the initial distribution collapses toward a lower dimensional repeller as the integration time grows. As the antisteady state is explored for longer times t, the initial state at $t = 0$ is constantly changing! The $t = 0$ distribution of states is therefore not a well-defined, fixed distribution! The only way the present state is well defined is if the *entire* future history of the Universe is known and fixed. Events that occur in the distant future influence the present distribution of states. If one includes the possibility of radioactive decay occurring in the future and influencing the present state, its innately random nature would make the present state of the Universe undefined. Likewise, the exercise of human free will at some time in the future could also change the present distribution of states! This would appear to be logically impossible in a universe where time increases rather than decreases.

The axiom of causality is so natural that people fail to observe that they have made this assumption. It is constantly used in classical mechanics. This is evidenced by the simple fact that Laplace transforms are defined only by $(0, \infty)$ time integrals rather than $(-\infty, \infty)$ time integrals as is used for spatial Fourier transforms. This, in turn, leads to *memory* functions rather than antimemory or forgetful functions. For an extensive discussion of causality and thermodynamics, see Evans and Searles (1996).

The transient fluctuation theorem and time-dependent response theory are meant to model the following types of experiment (Evans and Searles, 2002). One *begins* an experiment with a well-defined ensemble of systems characterized by some given *initial* (often equilibrium) distribution function. One *then* does something to the system (applies or turns off a field as the case may be), and one tries to predict what *subsequently* happens to the system. It is completely natural that one assumes that the probability of *subsequent* events can be predicted from the probabilities of finding *initial* phases and a knowledge of *preceding* changes in the applied field and environment of the system.

As we will soon see, computer simulation provides a clear illustration of the fact that the equations of motion can be run forward or backward. Those equations of motion are completely time-reversal-symmetric. It is the use of causality to predict the outcomes of experiments that actually breaks the symmetry of time.

Definition

The future state of the system is predicted solely from the probabilities of states of the system in the past. This is called the *axiom of causality*.

It is logically possible to compute the probability of occurrence of present states from the probabilities of future events, but this seems totally unnatural. A major problem with this approach is that, at any given instant, the future states are generally not known! In spite of these philosophical and practical difficulties, we will explore the logical consequences of the (unphysical) axiom of *anti-causality.*

We now show that if we derive the Green – Kubo relations for the transport coefficients defined by *anti-causal* constitutive relations, first, these anti-transport coefficients have the opposite sign to their causal counterparts and, second, what we call the system "response" starts to change *before* external fields are changed (Evans and Searles, 1996). In an anti-causal world, it becomes overwhelmingly probable to observe *final* equilibrium microstates that evolved from second law violating nonequilibrium steady states. Although this behavior is *not* seen in the macroscopic world, *anti-causal* behavior is permitted by the solution of the time-reversible equations of motion, and we demonstrate, using computer simulation, how to find phase space trajectories that exhibit *anti-causal* behavior.

9.2 Causal and Anti-causal Constitutive Relations (Evans and Searles, 1996, 2002)

Consider the component $dB(t_1)$ of the linear response at time $t_1$of a system characterized by a response function $L(t_1, t_2)$. The response is due to the application of an external force $F(t_2)$ acting for an infinitesimal time $dt_2(> 0)$, at time t_2, and could be written as

$$\delta B(t_1) = L(t_1, t_2)F(t_2)\delta t_2 \tag{9.1}$$

This is the most general linear scalar relation between the response and the force components. If the system is autonomous – independent of the time at which the experiment is undertaken (i.e., the same response is generated when both times appearing in Eq. (9.1) are translated by an amount $t : t_2 \to t_2 + t,\ \ t_1 \to t_1 + t$), then the response function $L(t_1, t_2)$ is solely a function of the difference between the times at which the force is applied and the response is monitored:

$$\delta B(t_1) = L(t_1 - t_2)F(t_2)\delta t_2 \tag{9.2}$$

Definition

The invariance of the response to time translation is called the *assumption of stationarity.*

Equation (9.2) does not, in fact, describe the results of actual experiments because it allows the response at time t_1 to be influenced not only by forces in the past, $F(t_2)$, where $t_2 < t_1$, but also by forces that have not yet been applied $t_2 > t_1$. We therefore distinguish between the causal and anti-causal response components

$$\delta B_C(t_1) \equiv +L_C(t_1 - t_2)F(t_2)\delta t_2, \quad t_1 > t_2 \tag{9.3a}$$

$$\delta B_A(t_1) \equiv -L_A(t_1 - t_2)F(t_2)\delta t_2, \quad t_1 < t_2 \tag{9.3b}$$

Later we will prove that $L_C(t) = L_A(-t)$.

(Note: we could consider superimposing a causal and an anti-causal response. However by considering these separately, we argue later that the anti-causal component leads the present state of the Universe undefined. So we simply argue that in our Universe with increasing time, anti-causal responses are physically impossible.)

Considering the response at time t to be a linear superposition of influences due to the external field at all possible previous (or future) times gives

$$B_C(t) = \int_{-\infty}^{t} L_C(t - t_1)F(t_1)dt_1 \tag{9.4a}$$

for the causal response and

$$B_A(t) = -\int_{t}^{+\infty} L_A(t - t_1)F(t_1)dt_1 \tag{9.4b}$$

for the anti-causal response.

9.3 Green–Kubo Relations for the Causal and Anti-causal Response Functions (Evans and Searles, 1996, 2002)

> And if also the materialistic hypothesis of life were true, living creatures would grow backwards, with conscious knowledge of the future, but no memory of the past.
>
> *(Thomson, 1874)*

To make this discussion more concrete, we will discuss the Green–Kubo relations for shear viscosity (Evans and Morriss, 1990). Analogous results can be derived for each of the Navier–Stokes transport coefficients. We assume that the regression of fluctuations in a system at equilibrium, whose constituent particles obey Newton's equations of motion, is governed by the Navier–Stokes equations.

Definition

We consider the *wave vector-dependent transverse momentum density* $J_\perp(k_y, t)$:

$$J_\perp(k_y, t) \equiv \sum_i p_{xi}(t)e^{ik_y y_i(t)} \tag{9.5}$$

where p_{xi} is the x-component of the momentum of particle i, y_i is the y-coordinate of particle i, and k_y is the y-component of the wave vector.

The (Newtonian) equations of motion can be used to calculate the rate of change of the transverse momentum density. They give

$$\begin{aligned} \dot{J}_\perp &= ik_y \left[\sum_i p_{xi}p_{yi}e^{ik_y y_i} + 1/2 \sum_{i,j} y_{ij}F_{xij} \frac{1 - e^{ik_y y_{ij}}}{ik_y y_{ij}} e^{ik_y y_i} \right] \\ &\equiv ik_y P_{yx}(k_y, t) \end{aligned} \tag{9.6}$$

In this equation, F_{xij} is the x-component of the force exerted on particle i by particle j, $y_{ij} \equiv y_j - y_i$, and $P_{xy}(k_y, t)$ is the xy-component of the wave vector-dependent pressure tensor. For simplicity, we assume the interparticle forces are simple pair interactions. For such systems, Eq. (9.6) is exact.

We now consider the response of a pressure tensor to a strain rate $\dot{\gamma}(t)$ applied to the fluid for $t > 0$ in the causal system and for $t < 0$ in the anti-causal system. In the causal system, the strain rate is turned on at $t = 0$, while in the anti-causal system the strain rate is turned off at $t = 0$. Since the pressure tensor is related to the time derivative of the transverse momentum current by Eq. (9.6) and the strain rate is related to the Fourier transform of the transverse momentum density by $\dot{\gamma}(k_y, t) = -ik_y J_\perp(k_y, t)/\rho$, the most general linear, stationary, and causal constitutive relation can be written as

$$\dot{J}_\perp(k_y, t) = \frac{-k_y^2}{\rho} \int_0^t \eta_C(k_y, t-s) J_\perp(k_y, s) ds, \quad t > 0 \tag{9.7}$$

where η_C is the causal response function (or memory function), and ρ is the mass density. The corresponding anti-causal relation is

$$\dot{J}_\perp(k_y, t) = \frac{k_y^2}{\rho} \int_t^0 \eta_A(k_y, t-s) J_\perp(k_y, s) ds, \quad t < 0 \tag{9.8}$$

where η_A is the anti-causal "response" function. Note that because $t < 0$, we find that the argument $(t - s)$ in Eq. (9.8) is less than zero, and we are indeed exploring the response of the system at times less than zero, which is prior to the changes in the strain rate that occur at times greater than zero!

It is straightforward to use standard techniques to evaluate the Green–Kubo relations for the causal and anti-causal shear viscosity coefficients.

Definitions

In the anti-causal case, it is important to remember that the usual *Laplace transform*

$$\widetilde{F}(s) \equiv \int_0^{+\infty} F(t) e^{-st} dt, \quad t \geq 0 \tag{9.9}$$

is inappropriate and needs to be replaced by an *anti-Laplace transform*

$$\widehat{F}(s) \equiv \int_{-\infty}^0 F(t) e^{st} dt, \quad t \leq 0 \tag{9.10}$$

(Note: $\widehat{F}(s) = \displaystyle\int_0^\infty F(-t) e^{-st} dt = \widetilde{F}'(s), \ \ t \geq 0$, where $F'(t) \equiv F(-t)$). We note that the anti-Laplace transform of a time derivative is $\hat{\dot{F}}(s) = F(0) - s\widehat{F}(s)$, and that the anti-Laplace transform of a convolution is the product of the anti-Laplace transforms of the convolutes.

By multiplying both sides of Eqs. (9.7) and (9.8) by $J_\perp(-k_y, 0)$ and taking the (equilibrium) ensemble average, one can easily derive the following relations for the shear viscosity and the anti-causal shear viscosity:

$$\widetilde{C}(k_y, s) = \frac{C(k_y, 0)}{s + \frac{k_y^2 \widetilde{\eta}_C(k_y, s)}{\rho}} \tag{9.11a}$$

$$\widehat{C}(k_y, s) = \frac{C(k_y, 0)}{s + \frac{k_y^2 \widehat{\eta}_A(k_y, s)}{\rho}} \tag{9.11b}$$

where

$$C(k_y, t) \equiv \langle J_\perp(k_y, t) J_\perp(-k_y, 0)\rangle, \quad \forall t \tag{9.12}$$

More useful relations for the viscosity coefficients, especially at $k = 0$, can be obtained by utilizing the equilibrium stress autocorrelation function

$$N(k_y, t) \equiv \frac{1}{V k_B T}\langle P_{yx}(k_y, t) P_{yx}(-k_y, 0)\rangle, \quad \forall t \tag{9.13}$$

Using the fact that $\widehat{N} = -\ddot{\widehat{C}}/k_y^2 V k_B T$, one can show (Evans and Morriss, 1990)

$$\widetilde{\eta}_C(k_y, s) = \frac{\widetilde{N}(k_y, s)}{1 - k_y^2 \widetilde{N}(k_y, s)/\rho s} \tag{9.14a}$$

$$\widehat{\eta}_A(k_y, s) = \frac{\widehat{N}(k_y, s)}{1 - k_y^2 \widehat{N}(k_y, s)/\rho s} \tag{9.14b}$$

At zero wave vector, we find that the causal and anti-causal memory functions are both given by the equilibrium autocorrelation function of the pressure tensor:

$$\begin{aligned} \eta_C(t) &= \eta_A(-t), \quad \text{where } t > 0 \\ &\equiv \eta(t), \quad \forall t \\ &= \frac{V}{k_B T}\langle P_{yx}(t) P_{yx}(0)\rangle \end{aligned} \tag{9.15}$$

where we have used $P_{yx}(t)V = \lim_{k \to 0} P_{yx}(k_y, t)$. Since equilibrium autocorrelation functions are symmetric in time, one does not have to distinguish between the positive and negative time domains. This proves our assertion made in Section 9.2 that $L_C(t) = L_A(-t)$.

Using Eqs. (6.13)–(6.15) and taking the zero wave vector limit, we obtain the causal response of the xy-component of the pressure tensor as

$$P_{yxC}(t) = -\int_0^t \eta(t - s)\dot{\gamma}(s) ds, \quad t > 0 \tag{9.16}$$

and the anti-causal response as

$$P_{yxA}(t) = \int_t^0 \eta(t - s)\dot{\gamma}(s) ds, \quad t < 0 \tag{9.17}$$

In the linear regime close to equilibrium, the instantaneous dissipation function $\Omega(t)$ for an isokinetic system is given by

$$\Omega(t) = -\beta P_{yx}(t)\dot{\gamma}(t)V \tag{9.18}$$

where $\dot{\gamma}(t)$ is the time-dependent strain rate. From Eqs. (9.16) and (9.17), it is easy to see that if we conduct two shearing experiments, one on a causal system with a strain rate history $\dot{\gamma}_C(t)$ and one on an anti-causal system with $\dot{\gamma}_A(t) = \pm\dot{\gamma}_C(-t)$, then

$$\Omega_A(t) = -\Omega_C(-t) \tag{9.19}$$

This proves that, if the causal system satisfies the second law of thermodynamics, then the anti-causal system must violate that law, and vice versa. If we now invoke the second law inequality, we see the following:

$$\int_0^t ds\langle\Omega_A(-s)\rangle = -\int_0^t ds\langle\Omega_C(s)\rangle \leq 0, \quad \forall t > 0 \tag{9.20}$$

Definition

Equation (9.20) is the *anti-second law inequality.*

9.4
Example: The Maxwell Model of Viscosity (Evans and Searles, 1996, 2002)

In this section we examine the consequences of the causal and anti-causal response by considering the Maxwell model for linear viscoelastic behavior (see Section 2.4 of Evans and Morriss (1990)). If we consider the causal response of a system to a two-step strain rate ramp

$$\begin{aligned} \dot{\gamma}_C(t) &= a, \quad 0 < t < t_1 \\ \dot{\gamma}_C(t) &= b, \quad t_1 < t < t_2 \end{aligned} \tag{9.21}$$

then use the Maxwell memory kernel

$$\eta_M(t) = G_\infty e^{-|t|/\tau_M}, \quad \forall t \tag{9.22}$$

in Eq. (9.16) and the fact that the causal η_C and anti-causal η_A Maxwell shear viscosities in the zero frequency limit are

$$\eta_C = \eta_A = G_\infty \tau_M = \eta_M \tag{9.23}$$

we find that the causal response is

$$\begin{aligned} P_{xyC}(t) &= -a\eta(1 - e^{-t/\tau_M}), \quad 0 < t < t_1 \\ P_{xyC}(t) &= -a\eta(e^{-(t-t_1)/\tau_M} - e^{-t/\tau_M}) - b\eta(1 - e^{-(t-t_2)/\tau_M}), \quad t_1 < t < t_2 \end{aligned} \tag{9.24}$$

If we now consider the corresponding anti-causal experiment with strain rate histories given by

$$\dot{\gamma}_A(t) = a, \quad -t_1 < t < 0$$
$$\dot{\gamma}_A(t) = b, \quad -t_2 < t < -t_1 \tag{9.25}$$

we find that the anti-causal response is

$$P_{xyA}(t) = a\eta(1 - e^{t/\tau_M}), \quad -t_1 < t < 0$$
$$P_{xyA}(t) = a\eta(e^{(t+t_1)/\tau_M} - e^{t/\tau_M}) + b\eta(1 - e^{(t+t_2)/\tau_M}), \quad -t_1 < t < -t_2 \tag{9.26}$$

From Eqs. (9.24) and (9.26), it is clear that

$$P_{xyA}(t) = -P_{xyC}(-t) \tag{9.27}$$

These response functions are shown graphically in Figure 9.1. A two-step strain rate ramp with $a = 1.0$, $b = 0.5$, $t_1 = 2$, and $t_2 = 4$ was considered. Equations (9.24) and (9.26) were used to predict the causal and anti-causal responses, respectively, of the xy-component of the pressure tensor. Values of $G_\infty = 40.0$ and $t = 0.05$ were used in the model. These values were obtained from approximate fits to computer simulation data (see Section 6.5).

The data in Figure 9.1 show that, for the causal response, P_{xy} is zero at equilibrium ($t \leq 0$) and decreases when the field is applied until the steady-state value is obtained. It remains at the steady value until $t = 2$, at which time the strain rate is reduced. Since this system is causal, no change in P_{xy} occurs until *after* the strain rate is reduced, when it increases until the system reaches a new steady state. We display the anti-causal response from $t = -4$ where it is in an antisteady state. Just *before* the strain rate is increased (at $t = -2$), P_{xy} increases to a new antisteady state value. Using Eq. (9.18), we see that in the causal response dissipation is in the graph always positive and second saw satisfying, whereas in the anti-causal response the instantaneous dissipation is in this graph always negative.

9.5
Phase Space Trajectories for Ergostatted Shear Flow (Evans and Searles, 1996, 2002)

We now examine the causal and anti-causal response on a microscopic scale, and we consider the relative probability of observing second law satisfying and second-law-violating trajectories by studying an ergostatted system of N particles under shear.

The ergostatted SLLOD equations of motion (2.21) and (2.23) are time-reversible (Evans and Morriss, 1990). Therefore, for every i-segment $S^t\boldsymbol{\Gamma}_{(i)}$, $(0 < t < \tau)$, there exists a conjugate trajectory segment $S^t\boldsymbol{\Gamma}_{(i^{(K)})}$, $(0 < t < \tau)$ with the property that $P_{xy}(S^t\boldsymbol{\Gamma}_{(i^{(K)})}) = -P_{xy}(S^{-t}\boldsymbol{\Gamma}_{(i)})$, $(0 < t < \tau)$. Thus, the t-averaged shear stress $\overline{P}_{xy,i,t} \equiv {}^1\!/_t \int_0^t ds P_{xy}(S^s\boldsymbol{\Gamma}_i)$ for segment i is equal and opposite to that for its conjugate: $\overline{P}_{xy,i^K,t} = -\overline{P}_{xy,i,t}$. We note that, since the solution of the

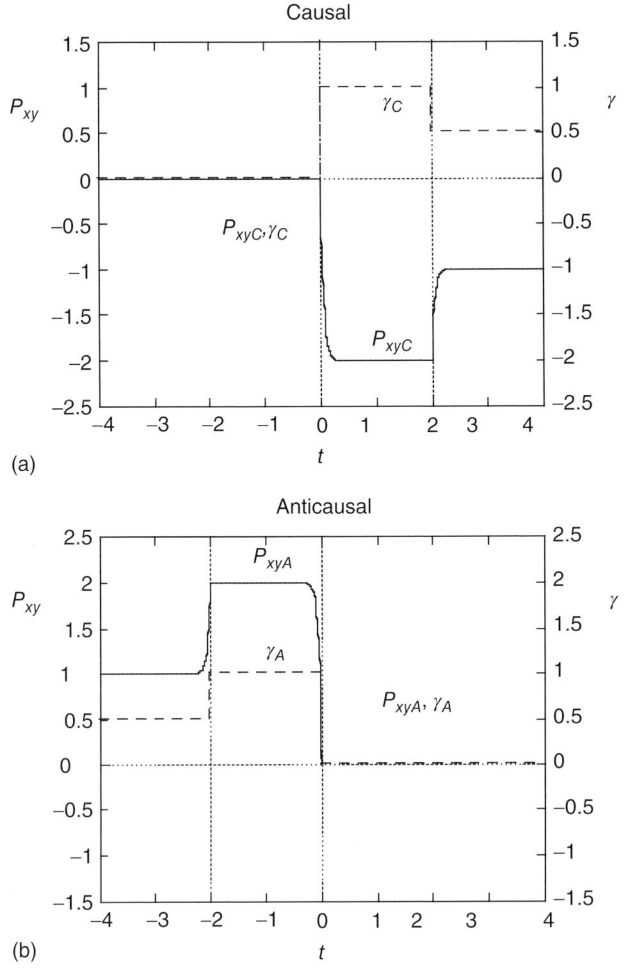

Figure 9.1 A schematic diagram of the (a) causal and (b) anti-causal response of P_{xy} to a two-step strain rate ramp determined using the Maxwell model for linear viscoelastic behavior with $G_\infty = 40$ and $\tau_M = 0.05$ (solid line). In both cases the time dependence of the strain rate is shown as a dashed line. Reproduced from Evans and Searles (1996) with permission of American Physical Society.

equations of motion is a unique function of the initial conditions, the conjugate segment is also unique.

We have previously shown that for shear flow, conjugate segments may be generated by using a phase space mapping known as a *Kawasaki- or K-map* see Section 7.4 of Evans and Morriss (1990) and section 2.3. A K-map of a phase $\mathbf{\Gamma}$ is defined as a time-reversal map that is followed by a y-reflection. In the case of shear flow, the K-map leaves the strain rate unchanged but changes the sign of the shear stress, that is, $M^K\mathbf{\Gamma} = M^K(x, y, z, p_x, p_y, p_z) = (x, -y, z, -p_x, p_y, -p_z) \equiv \mathbf{\Gamma}^{(K)}$.

It is straightforward to show that the Liouville operator for the system simulated by Eqs. (2.21) and (2.23), $iL(\mathbf{\Gamma}, \dot{\gamma}) \equiv \sum [\dot{\mathbf{q}}_i(\mathbf{\Gamma}, \gamma)\cdot\partial/\partial\mathbf{q}_i + \dot{\mathbf{p}}_i(\mathbf{\Gamma}, \dot{\gamma})\cdot\partial/\partial\mathbf{p}_i]$, has the property that under a K-map $M^{(K)}iL(\mathbf{\Gamma}, \dot{\gamma}) = iL(\mathbf{\Gamma}^{(K)}, \dot{\gamma}^{(K)}) = -iL(\mathbf{\Gamma}, \dot{\gamma})M^{(K)}$. If we assume a strain rate history such that, $\dot{\gamma}_K(-t) = \dot{\gamma}(t), \ \ \forall t$, then it follows that if a K-map is carried out on an arbitrary phase $\mathbf{\Gamma}$ at $t = 0$, then evolution *forward* in time from $\mathbf{\Gamma}^{(K)}$ under a strain rate $\dot{\gamma}_K(t)$ is equivalent to time evolution *backward* in time from $\mathbf{\Gamma}$ under the strain rate history $\dot{\gamma}(t)$, $(t < 0)$:

$$P_{xy}(-t, \mathbf{\Gamma}, \dot{\gamma}(-t)) = \exp[-iL(\mathbf{\Gamma}, \dot{\gamma}(-t))t]P_{xy}(\mathbf{\Gamma}) = -P_{xy}(t, \mathbf{\Gamma}^{(K)}, \dot{\gamma}_K(t)) \qquad (9.28)$$

We note that, if we do not assume that $\dot{\gamma}_K(-t) = \dot{\gamma}(t), \ \ \forall t$, then there is no general method for generating conjugate trajectory segments. This is because propagators with different strain rates do not commute and the inverse propagator must therefore retrace the strain rate history of the conjugate propagator but in inverse historical order.

We will now indicate, in more detail, how to construct the conjugate segment $i^{(K)}$, from an arbitrary phase space trajectory segment i. The construction is illustrated in Figure 9.2 for the case where the strain rate remains the same for the duration of the trajectory. A trajectory of length τ is generated by solving the equations of motion. The conjugate segment is then constructed by applying a K-map to the phase at the midpoint of the segment $(t = \tau/2)$, $M^K\mathbf{\Gamma}_{(2)} = \mathbf{\Gamma}_{(5)}$. We then

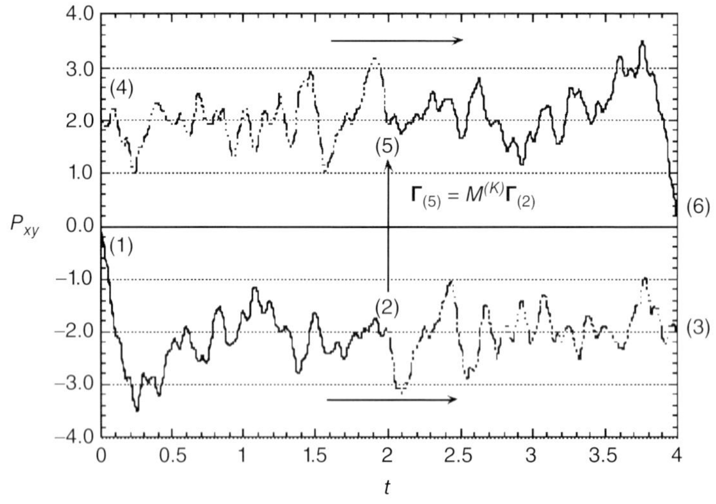

Figure 9.2 P_{xy} for trajectory segments from a simulation of 200 WCA disks at $T = 1.0$ and $n = 0.8$. A constant strain rate of $\gamma = 1.0$ is applied at $t = 0$. The trajectory segment $\mathbf{\Gamma}_{(1,3)}$ was obtained from a forward time simulation. At $t = 2$, a K-map was applied to $\mathbf{\Gamma}_{(2)}$ to give $\mathbf{\Gamma}_{(5)}$. Forward and reverse time simulations from this point give the trajectory segments $\mathbf{\Gamma}_{(5,6)}$ and $\mathbf{\Gamma}_{(5,4)}$, respectively. If one inverts P_{xy} in $P_{xy} = 0$ and inverts time about $t = 2$, one transforms the $P_{xy}(t)$ values for the antisegments $\mathbf{\Gamma}_{(4,6)}$ into those for the conjugate segment $\mathbf{\Gamma}_{(1,3)}$. Reproduced from Evans and Searles (1996) with permission of American Physical Society.

advance in time from the point ($\mathbf{\Gamma}_{(5)}$) to $t = \tau$ by solving the equations of motion, and also go backward in time from the K-mapped point $t = \tau/2$ to $t = 0$. A conjugate trajectory of length τ is thereby produced. This construction has previously been described in more detail (Evans and Searles, 1995).

Clearly, the mapped trajectory is a solution of the equations of motion for the system, and therefore it would eventually be observed from the ensemble of starting states. When the K-map is carried out at $t = 0$, the shear stress is inverted, and Eq. (6.18) shows that $P_{xy}(\tau/2 + t, \mathbf{\Gamma}) = -P_{xy}(\tau/2 - t, \mathbf{\Gamma}^{(K)})$, and similarly $P_{xy}(\tau/2 - t, \mathbf{\Gamma}) = -P_{xy}(\tau/2 + t, \mathbf{\Gamma}^{(K)})$; therefore, for every point on the original trajectory, there is a unique point on the mapped trajectory with opposite shear stress. The τ-averaged shear stress of the conjugate trajectory is opposite to that of the original trajectory, that is, $\overline{P}_{xy,i^K}(\tau) = -\overline{P}_{xy,i}(\tau)$. Thus, if the original segment was one satisfying the second law, then the conjugate segment is one violating the second law, and vice versa.

In a *causal* world, which is described by causal macroscopic constitutive relations such as Eq. (9.4a), the observed segments are overwhelmingly likely to be satisfying the second law. It is a simple matter to apply the arguments of Section 2.1 for the special case of ergostatted shear flow, where a simple time-reversal map cannot be used and must be replaced by the K-map. The condition of ergodic consistency has to be modified slightly to require

$$f(S^t\mathbf{\Gamma}^K; 0) \neq 0, \quad \forall \mathbf{\Gamma} \in D \tag{9.29}$$

The result is the ESFT given in Eq. (3.6).

9.6 Simulation Results (Evans and Searles, 1996, 2002)

We can demonstrate the relationships between the conjugate pairs of trajectories, the second law of thermodynamics, and causal and anti-causal response using numerical simulations of the system described by Eqs. (2.21) and (2.23). Figure 9.3 shows the response of P_{xy} for a trajectory and its conjugate when a constant strain rate is applied. The response was determined using nonequilibrium molecular dynamics simulations of 200 disks in two Cartesian dimensions. The disks interact via the WCA potential

$$\phi(r) = \begin{cases} 4\left(r^{-12} - r^{-6}\right) + 1 & r < 2^{1/6} \\ 0 & r > 2^{1/6} \end{cases} \tag{9.30}$$

Shearing periodic boundary conditions were used to minimize boundary effects (Evans and Morriss, 1990). The system was maintained at a constant kinetic temperature of $T = 1.0$, and the particle density was $n = N/V = 0.8$. An initial phase was selected from an equilibrium distribution, and a strain rate of $\gamma = 1.0$ was applied to the system at $t = 0$. A trajectory segment was generated by simulating forward in time to $t = 4$. The conjugate trajectory was constructed using the scheme describe above. Examination of the trajectories shows that $P_{xy}(\tau + t)$ for

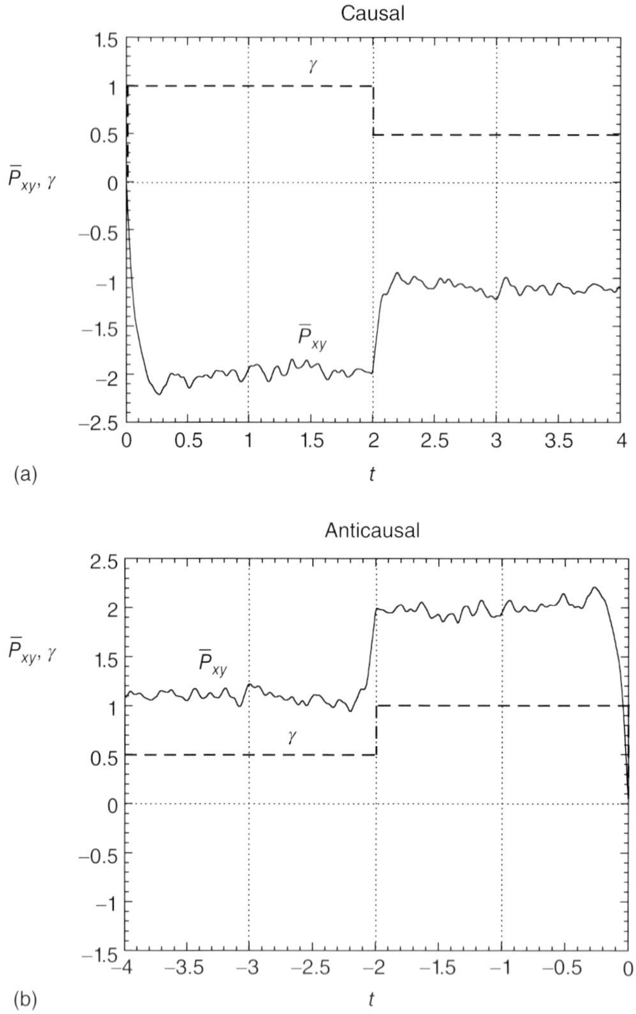

Figure 9.3 $\bar{P}_{xy}$ (solid line) from nonequilibrium molecular dynamics simulations of 56 particles at $T = 1.0$ and $n = 0.8$ undergoing shear flow. The dashed line gives the time dependence of the strain rate. In (a), $\bar{P}_{xy}$ was determined using 1000 trajectories whose initial phases were selected from an equilibrium distribution, and to which a two-step strain rate was applied. (b) Shows $\bar{P}_{xy}$ for their conjugate trajectories. The conjugate trajectories were obtained by applying a K-map to the phase of the trajectory at $t = 2$, simulating forward and backward in time from this point and translating in time so that the conjugate trajectory ends at $t = 0$. Note that the strain rate history of the conjugate trajectory is reversed. Reproduced from Evans and Searles (1996) with permission of American Physical Society.

the trajectory satisfying the second law is equal in magnitude but opposite in sign to $P_{xy}(\tau - t)$ for the trajectory violating the second law, where t is the time at which the K-map is applied ($\tau = 2$). These results therefore confirm the relationship between P_{xy} of trajectories satisfying the second law and the conjugate trajectories violating the second law, given by Eq. (9.28).

The causality of the response is more clearly demonstrated in Figure 9.3, where the response of P_{xy} to a strain rate ramp is shown. $P_{xy}(t)$ was averaged over 100 individual trajectories to reduce the fluctuations in the steady state, giving a partially ensemble-averaged response $\overline{P}_{xy}$. In these simulations, 56 disks were used. The initial phases of the trajectories shown in Figure 9.3 were sampled from the equilibrium distribution at $t = 0$. $\overline{P}_{xy}$ is close to zero at equilibrium and decreases to near a steady-state value after the field is applied. After the strain rate is reduced, $\overline{P}_{xy}$ increases toward a new steady-state value.

The conjugate trajectories are shown in Figure 9.3. They were constructed as described above and translated in time to begin at $t = -4$. At this time, the system is in an antisteady state and $\overline{P}_{xy}$ remains near its antisteady state value until just *before* the strain rate is changed, when it increases toward a new antisteady state value.

In accordance with the ESFT, these response curves demonstrate that most initial phases (here all 100 randomly selected initial phases) satisfy the second law and most phases (again all 100 initial random phases) exhibit response curves that we would describe as having "causal" characteristics (i.e., the stress responds to *prior* rather than *future* changes in the strain rate). Second law violating conjugate trajectories respond to the step in the strain rate *before* it is made, so they are *anti-causal.* Close inspection of the graph reveals that at all points along pairs of conjugate trajectories $P_{xy}(t)_{\text{trajectory}} = -P_{xy}(-t)_{\text{conjugate trajectory}}$, which follows from Eq. (9.28).

The system used in the simulations corresponds to that examined using the Maxwell model described in Section 9.4. Figure 9.3 shows the response, determined by nonequilibrium molecular dynamics simulation, to the same two-step strain rate ramp that was used to model the response shown in Figure 9.1. Comparison of these response curves indicates that the system is reasonably well represented by the Maxwell model.

We should also note that, if we generate an antitrajectory that has negative average dissipation, such a trajectory will not continue indefinitely. Because the sum of its Lyapunov exponents is positive while the sum of exponents for the trajectory is negative, the antitrajectory is less mechanically stable than its conjugate trajectory. Because no numerically computed trajectory is exact, this numerical error is amplified by the Lyapunov instability, and eventually the antitrajectory will decay into a trajectory with positive average dissipation.

If the error in any computed phase space position is δ, and if the particles have a dimensionless radius and average momentum of unity, the time required for the antitrajectory to decay is $-\ln(\delta)/\lambda_{\min}$, where $\lambda_{\min}$ is the smallest (i.e., the most negative) Lyapunov exponent for the trajectory with positive average dissipation. This decay has nothing to do with why the second "Law" is satisfied. The error

δ is not a material property. In an electrical circuit, the second "Law" is satisfied as soon as the voltage F_e is applied. In fact, the initial rate of increase in the electrical current density J is given by an equilibrium fluctuation formula that has nothing to do with noise, errors, or Lyapunov instability: $\lim_{t\to 0^+} d\langle J(t)\rangle/dt = -\beta V\langle J(0^-)^2\rangle_{\text{eq}}F_e$.

One might have thought that the instability of the antitrajectory would be very strong, and the slope $d\langle J\rangle/dt$, when the current crosses zero, would be very large. In fact, this is not so, and the crossing slope is typically much less than the initial slope caused by applying the voltage to the system which was at equilibrium at time zero! In a sense, then, the equilibrium state is more unstable under shear than is the antisteady state!

9.7 Summary and Conclusion

As we have seen throughout this book, it is dissipation and not phase space compression, entropy, or entropy production that features in the fluctuation, the dissipation, and the equilibrium relaxation theorems. Each of these theorems is exact for systems of arbitrary size and arbitrarily near to, or far from, equilibrium. It used to be said that for nonequilibrium systems virtually no exact results are known. That is most definitely not the case today.

At the end of this book, we are now in a position to identify the key quantity that facilitates the entire exposition. Dissipation dominates the theory. Although it was originally defined to give the probability ratios of observing in the same initial ensemble, sets of trajectories, and their conjugate sets of anti-trajectories, this definition (3.2) also involves a balance between energy change and phase space expansion or contraction. This is particularly obvious in equilibrating systems, Eq. (5.22). By losing a certain quantity of heat from an otherwise Hamiltonian system, the system also gives up a certain amount of phase space. The ratio of heat loss to phase space expansion is given by $k_B T$, the reciprocal of the integration factor for the heat appearing in the Clausius inequality for thermal reservoirs.

This quantity, $k_B T$, is the thermodynamic temperature of the underlying equilibrium state toward which the nonequilibrium system will relax, if it is so allowed. This is another key element of our theory.

The fluctuation theorems are proved by directly exploiting the time reversal symmetry of the dynamics. Time-reversed sets of trajectories and antitrajectories are actually exploited to prove the theorem. Indeed, in systems where these conjugate sets do not exist, the fluctuation relations are not valid – ergodic consistency has broken down. The theorems are so powerful and general, precisely because their proofs make so few assumptions.

In a Universe where time increases, it seems to us that causality is the only physical possibility available. In a Universe where time increases, the axiom of causality permits us to conduct experiments forward in time as indeed time increases. In an anti-causal Universe, we must know the future states before they are available, because time only increases. Worse still, the present state of the Universe is in fact *undefined* if future events contain random events as occur in many quantum processes. If time decreased rather than increased, an anti-causal Universe would be indistinguishable from our own, making the actual direction of time irrelevant.

The other feature of our thesis is the minor role played by entropy. Indeed, entropy was only mentioned for systems at equilibrium (Sections 5.7 and 8.5). Indeed, since Gibbs' second paradox was announced (that entropy is conserved by autonomous Hamiltonian dynamics), entropy has been problematic away from equilibrium. Our thesis shows that it is unnecessary to consider entropy, except for equilibrium systems where dissipation, on the other hand, is identically zero. Entropy and dissipation are thus seen to have perfectly complementary roles.

It seems astonishing that 150 years after Clausius made his famous remarks (Clausius, 1865)

> The energy of the Universe is constant.
> The entropy of the Universe tends to a maximum,

we have now come to such a different point of view. The ubiquity of Clausius' view is all the more astonishing, because of the logically correct criticisms of his arguments that were already made in the late nineteenth century (Bertrand, 1887) and very early in the twentieth century (Orr, 1904, 1905; Buckingham, 1905). These criticisms and others were summarized in the widely read encyclopedia article written by the Ehrenfests in 1912 (Ehrenfest and Ehrenfest, 1990). This article was translated into English in 1959. In the preface to the English translation, Tatiana Ehrenfest wrote: "At the time the article was written, most physicists were still under the spell of the derivation by Clausius of the second law of thermodynamics in the form of the existence of an integrating factor for the heat ... " Ehrenfest and Ehrenfest (1990, p. viii). Clausius's predictions were made based on equilibrium system arguments. Nonequilibrium systems, including the current state of our world, are not treated by Clausius. Because many systems can be treated as close to equilibrium, the results remain extremely useful, but the fact that they are not always valid is rarely recognized today.

Energy and the entropy are both constants for autonomous Hamiltonian dynamics but, on average, the time-integrated dissipation increases until at sufficiently late times in any isolated or thermostatted system it approaches a constant value. In that limit, the instantaneous dissipation is zero everywhere in the allowed phase space. This assumes, of course, that classical mechanics suffices to describe the dynamics of our system and that our system of interest is T-mixing. We know that both assumptions fail to hold for natural processes taking place across the natural Universe. However, to at last be able to understand the basic statistical thermodynamics of classical systems is a considerable improvement in our understanding of natural processes!

Unlike entropy and temperature of Clausius' classical thermodynamics, dissipation and the temperature of the underlying equilibrium state are, for the systems studied herein, always well defined, regardless of system size or the proximity of that system to equilibrium.

We have also at last come to realize the fundamental role played by causality in physics. Only the past influences the present. The future will, in turn, be influenced by the present. The future *cannot* influence the present. The equations of motion in physics are by themselves insufficient to predict what goes on in the Universe. Those equations must be supplemented with the axiom of causality. This axiom is so natural that most physicists, chemists, and engineers fail to realize that it is in fact an *assumption* and that an alternative mathematical possibility even exists. We have argued that although anti-causality is a mathematical possibility, it cannot be a physical possibility because it makes the present state of the Universe undefined when there are random quantum, or free-will, processes in the future.

The lack of recognition of the significance of causality is, however, precisely why the proof of the "laws" of thermodynamics had to wait so long.

References

Bertrand, J.L.F. (1887) *Thermodynamique*, Gauthier-Villars, Paris.

Buckingham, E. (1905) On certain difficulties which are encountered in the study of thermodynamics. *Philos. Mag.*, **9**, 208.

Clausius, R. (1865) Ueber Verschiedene Für Die Anwendungen Bequeme Formen Der Hauptgleichungen Der Mechanischen Wärmtheorie. *Ann. Phys. Chem.*, **125**, 353.

Ehrenfest, P. and Ehrenfest, T. (1990) *The Conceptual Foundations of the Statistical Approach to Statistical Mechanics*, Dover, Mineola, NY.

Evans, D.J. and Morriss, G.P. (1990) *Statistical Mechanics of Nonequilibrium Liquids*, Academic Press, London.

Evans, D.J. and Searles, D.J. (1994) Equilibrium microstates which generate 2nd law violating steady-states. *Phys. Rev. E*, **50**, 1645–1648.

Evans, D.J. and Searles, D.J. (1995) Steady states, invariant measures, and response theory. *Phys. Rev. E*, **52**, 5839–5848.

Evans, D.J. and Searles, D.J. (1996) Causality, response theory, and the second law of thermodynamics. *Phys. Rev. E*, **53**, 5808–5815.

Evans, D.J. and Searles, D.J. (2002) The fluctuation theorem. *Adv. Phys.*, **51**, 1529–1585.

Landau, L.D. and Lifshitz, E.M. (1969) *Statistical Physics*, Pergamon Press, Oxford.

Orr, W.M. (1904) On Clausius's theorem for irreversible cycles, and on the increase of entropy. *Philos. Mag. Ser. 6*, **8**, 509.

Orr, W.M. (1905) On Clausius' theorem for irreversible cycles, and on the increase of entropy. *Philos. Mag.*, **9**, 728.

Searles, D.J. and Evans, D.J. (1996) On the lifetimes of antisteady states. *Aust. J. Phys.*, **49**, 39–49.

Thomson, W. (1874) Kinetic theory of the dissipation of energy. *Nature*, **9**, 441.

Index